The Commercial Engineer

The Commercial Engineer

Tim Boyce

HAWKSMERE

© Timothy Boyce 1990

Published by Hawksmere Ltd
12–18 Grosvenor Gardens
London SW1W 0DH

British Library Cataloguing
in Publication Data
Boyce, Tim
    The commercial engineer.
1. Engineering industries. Management
I. Title
620'.0068

ISBN 1-85418-055-X

Production in association with
Book Production Consultants, Cambridge

Typeset by
KeyStar, St Ives, Cambridge

Printed in Great Britain by
Bookcraft (Bath) Ltd,
Midsomer Norton, Avon

This book is dedicated to my father

About the author

Tim Boyce is Commercial Manager at a major electronics
company and has also worked in government procurement.
He is an experienced lecturer in both the UK and US
and has written The Commercial Engineer from a highly
practical viewpoint describing underlying theory only to the
extent necessary to put practice into context.

Contents

④ The contract 33

Foreword

I have read with much interest Tim Boyce's book 'The Commercial Engineer' and I would recommend it to any engineer who wishes to be enlightened on the commercial aspects of his work.

Having lectured with Tim Boyce on this topic over a number of years I know how difficult it is to master matters such as fixed price incentive type contracts, express warranty, intellectual property rights, to say nothing of the subtleties of negotiating contracts.

This book makes very readable what can be a difficult and sometimes boring topic for engineers. It has been placed in the business context of today with entrance into the European Community fully in mind – all engineers may profit from what it presents.

Barry T Turner
Chartered Engineer

1

○ ○ ○ ○ ○ ○ C H A P T E R ○ ○ ○

Introduction

As suppliers of goods and services the objective is to provide the customer with the correct goods of the right quality at the right time.

Wrong!

The objective is to earn profit for the shareholders and on their behalf to sustain the company and facilitate its growth.

This simple misconception so frequently observed by the author amongst his colleagues in the engineering fraternity initiated the idea of this book.

All companies that are engaged in the business of designing and making products with the aim of selling them are obviously heavily dependent upon the quality and performance of their engineers. Throughout this book the expression 'engineer' is used generically to cover people of a technical background, whether in design, production, quality, estimating, etc., etc. The modern and successful company cannot afford to think of these vital assets as the 'backroom boys' or as people toiling away in the dark. It is fundamental that the engineer sees and acknowledges the corporate objectives and that he understands how best to make a valuable contribution beyond the strict confines of his own particular professional discipline.

The purpose of this book is to take the engineer on a tour of understanding that starts with how his company is created, through the nature of the contracts that underpin his existence and on to the areas in which he may make the broader contribution.

As in most matters, business success is a question of balance. The opening paragraph is not intended to say that the correctness, quality and timeliness of the goods should be sacrificed in the name of profit.

The intent is to put these aspects into perspective against the background of the primary objective. Indeed these considerations can be viewed as the means to the end. The linking between them is shown in Fig. 1.1.

Figure 1.1: Shareholders and customers – primary interactions

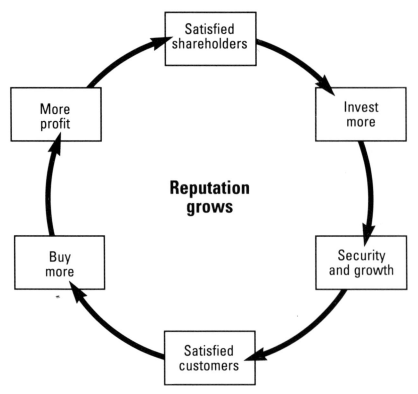

Successful business is self-perpetuating. Customers will return for more as orders are satisfied. Shareholders maintain or increase their investment as profits rise. Increasing orders and rising profits enhance reputation, encouraging both shareholders and customers to stay in this happy cycle.

Whilst the author's aim is to offer a broadening of outlook to the engineer, the aim throughout is to do this from a practical viewpoint based upon experience. All of the material included may be researched in considerable depth from a theoretical standpoint, but the intention here is to provide a jargon-free introduction.

2

● ● ● ● ● ● ● C H A P T E R ● ●

The business

① Introduction

The author found that, just as company employees often do not know the names or faces of their directors, it is usual for people not to understand the nature and structure of their company. An outline description serves to put the company into context against the environment within which it operates. This will be examined under the following headings:

- Classification
- Background
- Trading organisations
- Public and private companies
- Registered companies

② Classification

The most convenient way to introduce the method of constructing a business is firstly to explain the system of classifying trading organisations. That is to say, those organisations who under one form of authority or another are permitted under the law to carry on a trading activity.

It would be possible to classify trading organisations by an almost infinite number and permutation of parameters. For example, businesses could be sorted by nationality, product-range, number of employees, capital value and the like. It is usual, however, to classify by reference to their legal status, form of ownership and means of financing. The differing forms of trading organisation are shown in Table 2.1.

Table 2.1. Trading organisations

Primary classification
    Sole traders
    Partnerships (example of an unincorporated organisation)
    Joint stock company (example of an incorporated organisation)

Classification of corporations
    Incorporated by Royal Charter
    Incorporated by Parliament

Classification of Parliamentary Acts
    Special Act
    General Act

---

③ Background

In the UK by far the greatest number of businesses are carried on by sole traders, but by far the greatest volume and value of business is undertaken by joint stock companies and by public corporations. As government policy continues to take effect, more public corporations are taken into private ownership. Such organisations become joint stock companies and so the proportion by value of business undertaken by sole traders remains unchanged. In addition to sole traders and joint stock companies, the partnership is also a significant form of trading organisation. These three types between them are the ones of most interest in commercial life. The law and regulations surrounding each of these is vast and complex. However, it is possible to compare the main advantages and disadvantages of each, and these are illustrated in Figs. 2.1, 2.2 and 2.3. In the illustration shown, the joint stock company is of the type known as being incorporated with limited liability. This means that at the outset the shareholders or owners limit their liability to creditors in the event that the business becomes bankrupt.

Of the differing advantages between the three types of organisation, of particular interest is the question of separate legal existence. The joint stock company is recognised by the law as having a separate existence and thus it is the company and not its members who are liable for its debts and other legal obligations. The sole trader is in exactly the reverse position. The partnership (as an example of an unincorporated association) lives somewhere between the two positions. Strictly it has no separate legal existence but in some circumstances it can, for example, sue and be sued.

Figs 2.1, 2.2 and 2.3 show the shifting balance of advantage between the sole trader, partnership and limited-liability joint stock company. The sole trader is commercially the most vulnerable, as for him most things can go wrong.

Figure 2.1: Advantages and disadvantages of the sole trader

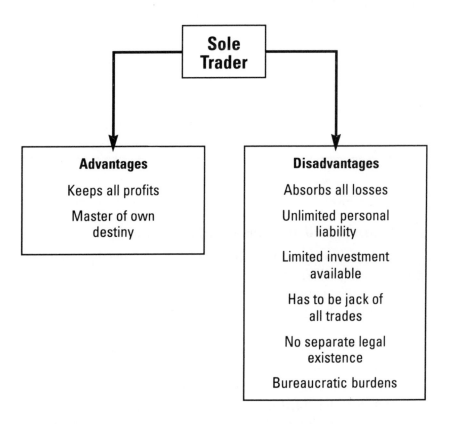

Figure 2.2: Advantages and disadvantages of partnership

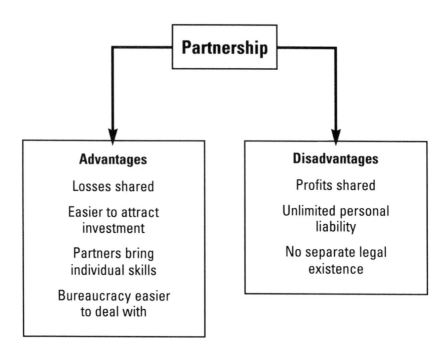

④ Public and private companies

There is an important distinction in UK law between public and private companies.

A public company has the following features:

a) It is limited by shares or guarantee.

b) It must have a minimum authorised shared capital.

c) It must possess a memorandum saying that it is a public company.

d) It must be registered under the Companies Act.

e) It must have a minimum of two members.

f) It may invite the general public to subscribe for its shares.

A private company is a company that does not satisfy all

Figure 2.3: Advantages and disadvantages of the joint stock company

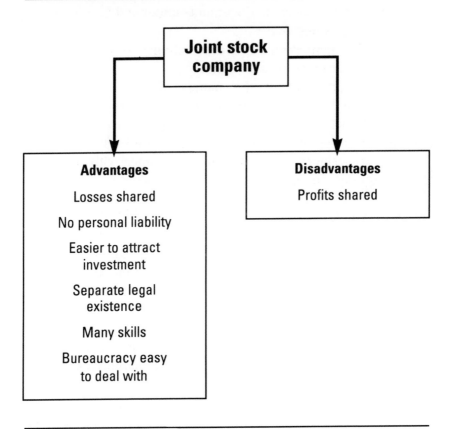

of (a)–(e) and it is a criminal offence for a private company to offer shares to the general public.

A private company may commence trading as soon as its certificate of incorporation under the Companies Act has been granted. The public company must wait until the Companies Registrar has satisfied himself and issued a certificate to the effect that the share capital requirements have been complied with.

Unlike the public company, the private company does not have to hold annual meetings. Both, however, must submit annual accounts to the registrar.

⑤ Registered companies

A registered company is created by submitting to the Registrar of Companies the following documents under and in compliance with the Companies Act:

a) The memorandum of association.

b) The articles of association.

c) A statement of the names of the intended first director(s) and the first secretary, together with their written consents to act as such. The statement must also contain the intended address of the company's registered office.

d) A statutory declaration of compliance with the Companies Act regarding registration.

e) A statement of the company's capital, unless it is to have no share capital.

The Registrar, when satisfied that the requirements have been complied with, issues a Certificate of Incorporation, which brings the company into legal existence. The memorandum of association and the articles of association are most important.

The Memorandum of Association

a) regulates the company's external affairs,

b) specifies the company name, objectives and powers,

c) indicates whether the members' liability is limited or not,

d) specifies the authorised share capital.

Thus from the memorandum of association investors can discover the purposes to which their money can be put. Also persons dealing with the company can discover vital information as to the size and powers of the company.

If the company is private the name must indicate the letters Ltd. If public, the letters plc must be included.

The articles of association in contrast are

a) to regulate internal administration,

b) to define the relationship between company and members,

c) to define the relationship between the members.

The chief officers of the company are its directors and company secretary.

The powers and responsibilities of the directors are

Table 2.2. Trading positions

| | Sole trader | Unincor-porated association | Joint stock company | |
			Public	Private
Legal status	Same as individual	*Not* a legal entity	A legal entity	A legal entiry
Subject to the Companies Act	Yes	Yes	Yes	Yes
Ownership of Property	Same as individual	Jointly by members	By the company	By the company
Liability for debts rests with	The trader	The members	The directors and members	The directors and members
Extent of liability	Unlimited	Usually limited	Usually limited	Usually limited
Registration required	No	No	Yes	Yes
Shares offered to public	No	No	Yes	No
Business commences	As trade commences	As trade commences	Grant of certificate that share capital requirements have been met	Grant of certificate of incorporation
Annual meeting	Not applicable	Optional	Statutory	Optional
Annual accounts	No	No	Yes (unlimited companies exempt)	Yes (unlimited companies exempt)

defined in the articles of association. The managing director will usually be given specific authority to

a) make certain decisions without reference to the board,
b) place or accept contracts,
c) authorise capital purchases.

The company secretary is usually appointed by the board of directors. The Companies Act states:

'it is the duty of the directors of a public company to take all reasonable steps to secure that the secretary is a person who appears to them to have the requisite knowledge and experience to discharge the functions of company secretary'.

In practice, the secretary will have to have appropriate experience, qualifications and hold membership of relevant professional institutes.

The salient points of this discussion of business construction are summarised and illustrated in Table 2.2.

3

⬤ ⬤ ⬤ ⬤ ⬤ ⬤ ⬤ **C** **H** **A** **P** **T** **E** **R** ⬤ ⬤

Commercial awareness

① Introduction

The key to success must be in the acquisition of commercial awareness. This phrase might be defined as the ability to take and understand the broad view across all of the objectives and operations of the business.

The six key words that encapsulate the concept of commercial awareness are:

- Profit
- Cash
- Orders
- Ideas
- Risk
- Contracts

The first four of these can be defined by reference to their immediacy of need:

The Now Needs: orders and cash

The Medium Need: profit

The Future Need: ideas

The company must have orders now to fill the factory, to produce deliveries and so to generate cash and profit. The company must have cash now to pay employees, suppliers, etc. The company must produce good profits at six- or twelve-monthly intervals. Poor profits means reduced capital investments by the company and loss of confidence by shareholders, who expect their bonuses. The company must generate new ideas, designs, products which will provide the orders, cash and profit in future times.

The final two topics may be considered as having major *influence* on profit, cash, orders and ideas. Risk management is crucial if these are not to be jeopardised. Similarly, the opportunity to maximise profit and cash, to eliminate threats to the successful performance of orders, and to protect ideas for the future hinges on the terms of the contract. The contract must be good and sound and familiar to all those concerned with it. There is an old maxim to the effect that if the job is going well the contract is left to gather dust in a forgotten drawer. If the job goes badly the contract does nothing to help.

In enshrining two extremes, the maxim neglects the majority of contracts, which lie somewhere between the two, and indeed in practice many problems are caused by people *thinking* they know what is in the contract without bothering to check. Also, if problems do arise then a solution which is arrived at other than by recourse to the courts will be based around what the contract says, even if the contract does not expressly address the specific problem.

② Maximising profits

If for these purposes profit is thought to be the difference between the cost of doing a job and the price that the customer pays, then clearly maximising profit means minimising cost, maximising selling prices and finding a product/market where customers abound. The well-tried policies towards this are illustrated in Fig 3.1. These are:

a) Concentration on producing and supplying those goods and services where demand is increasing. This may sound obvious but too frequently product developments are pursued because of the inherent technical challenge – found attractive at the individual or corporate level – rather than because of a well-researched market demand.

b) Minimisation of the cost of production by selecting the cheapest possible combination of premises, machinery and labour to use. Thus, if substituting machines for labour is cheaper, this will be done despite the social consequences of redundancies, etc.

c) Maintaining output at the level at which profits are maximised. For example, if the business is one of producing spoons and the best machine runs most efficiently at 5,000 spoons per hour then producing only 4,000 spoons per hour is under-utilising the equipment and recovering the cost of the machine more slowly – to attempt to produce 6,000 spoons an hour may incur higher maintenance costs and longer down-time periods.

d) Where a single organisation is dominant in its own area of activity it can affect the price of the goods it produces by varying the amount it supplies to the market. It is therefore able to adjust either price or output to suit its own profit-maximisation objectives.

Figure 3.1: Maximising profit

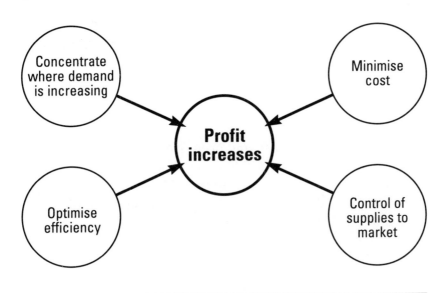

(a) and (d) are two sides of the same coin. (a) is to recognise that activity should take cognisance of the market and be influenced by it. (d) is to say that in favourable conditions it is possible to influence the market.

③ Cash

The matching income to expenditure, to yield a net positive or neutral position, is of paramount importance. If the net position is negative the firm will be unable to pay suppliers and employees, be unable to maintain interest payments on loan capital and may face bankruptcy. On the other hand, a net positive position is cash in hand to earn interest. There are many factors affecting cash-flow, including:

    a) Employee terms, e.g. monthly or weekly payments.

    b) Payment performance from customers where invoices are valid and payment due.

    c) Payment practices with suppliers; number of days credit.

    d) Timing of loan repayments and interest payments.

    e) Customer/supplier payment terms, e.g. payment on delivery, installation, down-payments or stage payments.

    f) Achieving contractual delivery schedules.

    g) Maintaining forecast production rates and schedules.

    h) Achieving standards to reduce inspection and test failures.

④ Orders

As well as generating cash and profit from performing existing orders, many companies will have a deliberate policy of growth and expansion. This necessarily may be accomplished at the short-term cost of reduced profit, as profits must be ploughed back into the business as a source of expansion financing. The possibilities for developing the business as shown in Fig 3.2 and include the following:

    a) Expanding existing markets, This may be achieved by developing new products and by increasing output combined with greater advertising, publicity and marketing activities.

    b) Successfully performing existing contracts. There is considerable value in terms of reputation in being seen to complete contracts on time and on performance. Goodwill is established and contented customers will come back for more. Every opportunity should be taken to let the customer know what a good deal he has had from you as he may not know himself.

c) Licensing ideas, manufacturing rights, dealership, agency and franchise rights to and/or from other companies.

d) Diversifying, i.e. by extending the product range or activity into new areas.

e) Takeover of other business organisations.

f) Merger/collaborate with other businesses.

Figure 3.2: Growth strategies

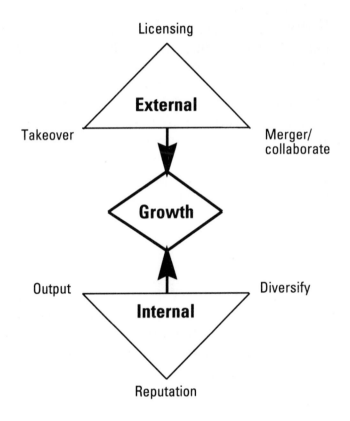

All well-run businesses will have defined plans for growth and expansion. Individuals will be responsible for drawing up specific plans and programmes for existing and new products and technologies. This will be geared towards existing and potential new customers and markets. The single most important thing is for every employee to keep in mind these fundamental growth strategies. Ideas should be put forward and every opportunity taken in talking to and meeting customers, suppliers, associates and colleagues. All contacts are people and all people like to sound knowledgeable and informed about their organisation. All intelligence gained needs to be shared to discover and exploit its value. Even if ostensibly worthless it may nevertheless corroborate, confirm or deny some other information.

⑤ Ideas

The company's ideas have a significant commercial value and the company owns the ideas and has the right to exploit them. Generically the ideas are known as 'intellectual property' and the rights to protect and exploit them as 'intellectual property rights'. 'Ideas' as a word does not quite go wide enough, as the intellectual property can, at the one extreme, be as simple as an approved customer list, and, at the other extreme, an invention to control the temperature of a nuclear reactor.

In any event, this intellectual property is lifeblood to the company and its future markets and products. Chapter 8 looks at this in some detail.

⑥ Minimising risk

Aiming for high profits is OK, but at what risk? It is no good pursuing that last pound if subsequently the business is unable to carry on. That is, the very process of carrying on business exposes the risk of commercial, financial or legal damage. So what are the risks and how can they be bounded? These exposures can be categorised broadly into:

    a) Technical
    b) Financial
    c) Commercial
    d) Legal
    e) Personnel
    f) Political
    g) Acts of God

These are shown in Fig 3.3.

Figure 3.3: Risk

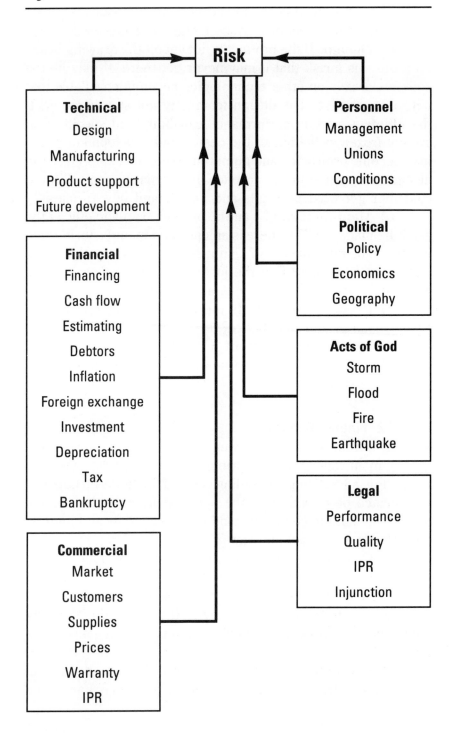

6.1. Technical risks

The word 'technical' is used here very loosely as a convenient heading. Technical risks include the following:

a) Design. If the product is not yet off the drawing-board then there is a risk that it may never get there. It may be too complex or rely on other technologies becoming available that fail to materialise. The design team may not stay together. If the ultimate product performance capability and specification are not frozen at the outset the design and development phase may become protracted and numerous tangents explored as the designers struggle in the absence of a concrete goal. Will the customer get what he wants and understand how to use it?

b) Manufacturing. Can the product be made efficiently; can it be made at all? If there is an optimum manufacturing rate can the shop-floor handle it? Can it be fitted in with other work and other schedules? Can the inspection and test procedures adequately satisfy quality needs and standards?

c) Product support. Is the product reliable; can it be easily repaired and maintained?

d) Future development Can the design be steadily updated and improved to meet new or revised requirements at minimum development cost.

6.2. Financial risks

a) Financing. Can the company find the funds for design, development and manufacturing? Do the funds exist for marketing, advertising and selling costs. If the funds do not exist can finance be raised? Will the bank or other lenders be co-operative? Can the company stand the repayments and interest until the product pays its way?

b) Cash flow. Once the product is into production can sales and output match one another, i.e. can the in-flow of receipts from customers match the out-flow of expenditure on materials, wages and operating costs? If not, can the company afford the cost of raising capital to finance this negative cash-flow?

c) Estimating. Do the actual costs of design and manufacture match those estimated? A selling price may have to be advertised before the product is ever made? Under-estimating could financially be a disaster.

d) Debtors. Will the customer pay on time or will he delay? Will he pay at all? Can the company afford the delays or the costs – legal fees and so on – of debt collection?

e) Inflation. The company will of course be affected by the general rise in prices and the cost of living.

f) Foreign exchange. Goods bought or sold in foreign currency will be at the risk of fluctuations in exchange rates.

g) Investments. The company's own financial investments will be exposed to the effect of money markets, stock-exchange confidence, international interest rates, etc.

h) Depreciation. The value of plant and equipment will decrease over the years.

i) Tax. Will the company have sufficient monies available when tax liabilities become due?

j) Bankruptcy/insolvency. If the financial performance fails, then ultimately the company may become bankrupt.

6.3. Commercial risks

a) Market. Does the market exist for the product? Are there sufficient customers? Is it a static or fluctuating market?

b) Customers. Are there diverse customers? Single customers or single-territory customers can dry up or become less in need. Demand can disappear or reduce earlier than expected.

c) Competitors. How strong is the competition? Will the competition have increased by the time the product is ready? Have the competitors' products and marketing strategy been analysed.

d) Suppliers. How safe are suppliers? Are there several sources of supply for each material requirement? Can single suppliers put you over a barrel? Are they committed to supply you for sufficient time? What influences are there over prices?

e) Prices. Is your selling price competitive but nevertheless sufficient to recover your costs and leave something for profit? Can the price be discounted for favoured customers? Can favourable payment terms be offered or secured?

f) Warranty. Will warranty claims be excessive and therefore costly?

g) Intellectual property. Will marketing the product enable competitors to make use of your ideas and inventions?

6.4. Legal risks

a) Non-performance. If you fail to meet, or are late in meeting a customer's order you are exposed to the possibility of legal action and the payment of damages.

b) Poor quality. If the goods do not fully meet all the customer's requirements you are exposed to legal action for breach of contract, breach of the Sale of Goods Act and/or the Misrepresentation Act. If, as a result of poor quality, property is damaged by the goods or people are injured or killed you may be sued for product liability under the Consumer Protection Act or otherwise pursued under the common law tort of negligence.

c) Intellectual property. If, accidentally or deliberately, you make use of somebody else's ideas you may be sued for infringement of patents, registered designs or copyright. Through careless marketing you may be infringing trade marks, trade names or goodwill.

d) Injunction. As a result of one or more of the foregoing a court may award an injunction against you, temporarily or permanently preventing you from carrying on with the business.

6.5. Personnel risks

a) Management. The management system may be deficient in structure or lacking in authority. Managers may not possess the requisite skills or mix of skills. Leadership may be lacking.

b) Trade Unions. Business operation may be disrupted by union activities. Official or unofficial strikes and other industrial action can disrupt manufacturing and output.

c) Conditions. In difficult times wage and salary levels may be low, other facilities (e.g. canteen and recreational) may be poor, leading to low morale and reducing efficiency and loyalty.

6.6. Political risks

a) Policy. As governments change and economic factors alter, so the emphasis, in terms of government support, on different technology sectors will change. Conservative governments traditionally support defence industries, for example, whilst Labour governments conventionally support mining and other heavy industries.

b) Economics. Again, as governments come and go, so the level and nature of industrial support varies. On one occasion tax concessions and support for exporters may be the order of the day. Times change and profit (any profit) is seen to be unpalatable and business has to suffer the consequences.

c) Geographical. Governments have for a long time offered financial incentives to industry in certain areas of high unemployment and these incentives and the geographical boundaries and locations may change with the evolution of government policy.

6.7. Acts of God
This includes storm, flood, lightening, earthquakes, etc.

6.8. Risk analysis
The categories of risk might be listed under the heading of those inside and those outside the company's control (Table 3.1).

Table 3.1.

Within the company's control	Outside the company's control
Technical	Government policies
Design	Tax
Manufacturing	Policy
Product support	Economics
Future development	Geography
Financial	Acts of God
Financing	Storm
Cash flow	Flood
Estimating	Earthquake
Debtors*	
Inflation*	
Foreign exchange*	
Investments*	
Depreciation	

Within the company's control (cont'd.)

Commercial
  Market*
  Customer*
  Competitors*
  Suppliers*
  Prices
  Warranty
  Intellectual property

Legal
  Non-performance
  Quality

Personnel
  Management
  Trade unions*
  Working conditions

These listings indicate that, with the exception of Acts of God or government, all risks are within the control of the company. This may seem to be a little harsh – clearly the company has no direct influence over the market-place and the competitors. (Nobody can guess at the outset whether a supplier will keep to his promise or whether the customer will pay on time). So perhaps it would be more fair to note that some – those marked with an asterisk – are shared between the two lists. The market-place is a volatile place and its vagaries are outside the company's direct control. However, within the company's control is the choice as to which sector of the market to attack and the manner in which that market may be addressed.

Another, more useful, way to segregate that elements in this list of risks is to differentiate between the ways in which they can be dealt with. The four categories are as follows:

a) Risks within the company's management responsibility.

b) Risks that the company may seek to pass on to the customer under the conditions of the contract.

c) Risks against which the company may try to buy insurance.

d) Risks totally outside the company's ability to manage or to protect itself.

In practice, some risks may lay across more than one of these categories.

In the area of technical risk the risks are largely within the gift of the company to manage: (1) by acquiring the right level and mix of skills the risk of poor design or design being delayed or not completed can be reduced; (2) by providing the right design aids and working environment the design process may be facilitated; (3) by freezing the ultimate requirement specification before substantial design gets under way. However, where the design objective is highly inventive or novel or where it is to meet special and unique requirements of a particular customer, it may be possible to pass some of the design risk on to the customer by phrasing the contract so that the work is not to meet the end-requirement but simply to work towards that goal to a predetermined level of expenditure. In this the customer takes the chance of the requirement not being met unless he chooses to spend more money. Similarly, in manufacturing, risk minimisation is largely attainable by good planning, good shop-floor layout, industrial engineering, etc. The customer may wish to share in the risk by providing funds towards the setting up or enhancement of the production facilities. High reliability, low maintenance, potential for further development can all be attained if included as design requirements at the outset.

Financial risks are also a combination of company management responsibility and a degree of passing the risk on to the customer. The company will be set up in the first place on certain assumptions regarding cash flow and this risk is within the management controls of the business.

Again, the customer may be persuaded to contribute to the alleviation of this problem by providing interim financing of high-value and/or long-duration contracts. Where there is a substantial foreign-currency content of the price, the customer may be content to make payments in whole or in part in the foreign currency, particularly if, as in the case of very large buyers of government departments, foreign currency can be bought more easily or more cheaply than by the supplier. Foreign-exchange risks and/or general inflation risks may be shared with the customer on the basis of a formula allowing for the price to be varied in line with movements in exchange rates or inflation rates. Estimating is largely a function of the skill of the engineers and surveyors, etc., responsible for estimating the cost of under-taking a job. Although matters such as tax liabilities and government financial support to industry are outside the scope of company management *per se*, nevertheless expert advice is available from bankers, finance consultants, and so on.

Commercial risks are essentially those flowing from, or associated with, the very nature of the particular enterprise or business. Thus choice of products, markets, customers and suppliers is a question of being prudent and exercising judgement and analysis. Monitoring market trends, the business columns and keeping watch for emerging competitors and products are all vital aspects. Taking steps to safeguard and exploit ideas and inventions are essential. Commercial risks are the very essence of being in business and they are almost entirely a subject for good management.

Legal risks cover more than one category. It is possible, for example, to pass on to the customer the risk of a third party taking action in respect of an infringement of a patent or copy-right. On the other hand, the company can actually take steps to ensure that it is not infringing such things by undertaking, for example, a patent search at the London Patent Office. Then again, it may be possible to insure against the risk of third-party legal action. Some legal requirements are assumed by the very act of carrying on a business. For example, the employer has a duty of care (dereliction of which is negligence) to employ-ees and customers. Goods offered for sale must actually be of merchantable quality.

Personnel risks are a matter for the company's own management. The employment of people is covered by a tre-mendous wealth of law and legislation. However, the risks to the success of the business are for the company to control and bound by good organisation, structure and management.

Political risks are outside the control of the company and

generally not insurable. The only real answer is to be dynamic and flexible enough to respond quickly and effectively to political changes that influence business.

Various aspects of risk and the considerations associated therewith are dealt with in more detail in later sections.

⑦ Contracts

Maximising profit, minimising risk and business development are all concerned with the carrying on of the business to the greater glory and profit of the shareholders. Nevertheless, this is achieved through the medium of winning contracts, on one end of which is the customer who is only concerned with acquiring the goods or services he needs at minimum cost and inconvenience. The important point is that under the contract the customer may be entitled to many things and all those involved in implementing a contract should be aware of those things and more importantly the extent and limitation of whatever the customer's rights are. Subsequent chapters go into this in some detail. The principle, however, is to know the customer's minimum and maximum entitlement. If in practice he is satisfied with his minimum entitlement then contract costs will have been minimised and profit maximised. On the other hand, if he is dissatisfied with his maximum entitlement then it needs to be put to him, carefully and diplomatically, that if he wants more, he must pay more.

⑧ Business analyses

To be set up in the first place the business needs money. This is either share capital (which is the capital belonging to the person setting up the business or the contribution made by others) or loan capital (which is money borrowed from a bank or other lender). Profits generated by successful performance can be shared between the investors and/or put back into the business to generate growth and development. This latter use is referred to as reserves. Thus financing the business derives from share capital, loan capital or reserves.

The business's expenditure in establishing the factory will be land, buildings, plant, machinery which will be kept, and these are referred to as fixed assets. Expenditure on materials, wages and salaries is known as working capital. If an element of earned profit is retained other than for reinvestment, perhaps for saving in a bank or investment in shares, etc, then this is known as investment.

The set-off of expenditure against financing is displayed in a company document called the balance-sheet. In a simple example the basic balance-sheet will therefore look like Table 3.2.

Table 3.2.

Expenditure	Financed by		
Fixed Assets	10,000	Share Capital	12,000
Working Capital	15,000	Loan Capital	12,000
Investments	5,000	Reserves	6,000
	£30,000		£30,000

The balance-sheet is there to effectively provide a snapshot in time of the location and accessibility of the company's money.

The important aspect about the balance-sheet is that it is a snapshot in time. The annual picture of the company's performance is found in the profit-and-loss account. The balance-sheet tells the company where its money is, and how easy it is to get hold of the money. The profit-and-loss account indicates the profit or loss generated by the firm's activities. A simple 'P+L' account might take the form shown in Table 3.3.

Table 3.3.

Sales	£50,000	
Interest on investments	£5,000	
Revaluation of propery	£10,000	£65,000
Less		
Materials	£20,000	
Labour	£10,000	
Overheads	£10,000	
Depreciation of fixed assets	£5,000	£45,000
	Net profit	£20,000

The importance of cash as the medium through which the business breathes has already been mentioned. All companies run an analysis of cash-flow to predict receipts against disbursements. The analysis can be done daily or monthly (or any periodicity), depending on the volatility and size of the company's operation.

The final principle analysis is the forecast order-intake/delivery plan. Quite simply, to sustain the operation on its present basis requires equivalent new orders to be won as existing orders are performed and completed. If the firm's optimum performance is in delivering 1,000 widgets a month and the average customer orders 100 at a time for delivery the same month, then the firm needs to secure 10 orders a month. Ideally, order-intake should run slightly ahead of deliveries. This provides some protection against problems and permits the steady growth of the company.

If order-intake runs well ahead of delivery then there may be the danger that the firm cannot cope with its commitments, or may run into cash-flow problems if raw material stocks have to be built up or further capital equipment bought ahead of receiving any payments from customers. On the other hand, if orders are trailing deliveries, then the level of output and performance for which the firm is currently geared will be too high for the falling orders and output will have to be reduced, and if the level of order-intake is not increased this contraction can become an interminable process to the point where the firm can no longer continue trading. One key point related to the order-intake/deliveries analysis is that, if orders begin to trail deliveries, then natural inertia, amongst other things, is one reason that manufacturing output will not reduce in step. A consequence of this is that the quantity of stock may build up. Stock is a current asset within working capital but the value of that stock is related to its market value. If the market, or the companies share of it, is contracting then the value of the stock can rapidly fall. Even in a healthy business the general maxim is to keep stock to the absolute minimum consistent with meeting the most likely forecast demand.

The measure of the problem can be seen by observing the change in net order-book value. The gross order-book is the summation of all gross contract prices. The net order-book value is the gross order-book value minus the value of deliveries already made under the orders held. A company delivering orders at a value of £30,000 per month against a gross order-book of £300,000 has 10 months work. If the net order-book value declines then trouble lies ahead, reinforcing the point made above. The word 'deliveries' (sometimes referred to as output) in this context is used in the sense of phased performance of the contract rather than necessarily the physical supply of goods. For example, if the nature of the business is the supply of consultancy services, then an annual contract of £12,000 in value could be considered as generating deliveries of £1,000 per month for the purposes of assessing net order-book.

In summary, the four important devices are:

a) Balance sheet. A snapshot of the location and accessibility of cash.

b) Profit-and-loss account. An annual picture of twelve-months performance.

c) Cash flow analysis. A (usually) monthly analysis of the financial health of the operation.

d) Order-intake deliveries analysis. A projection of the growth or contraction of the business.

These devices provide the major insight into the state of the business in so far as the present position is concerned. The future is considered in the plans that the business makes in terms of new products, new markets, new customers, new technologies and new capital investment. To understand and keep in mind these matters, together with the maximum-profit minimum-risk philosophy and an appreciation of contractual obligations, is to take on a broader view of the commercial side of life. Seeing the business's objectives, understanding its operations, knowing its strengths and weaknesses is crucial to gaining a full sense of commercial awareness (see Fig 3.4.).

Figure 3.4: Commercial awareness

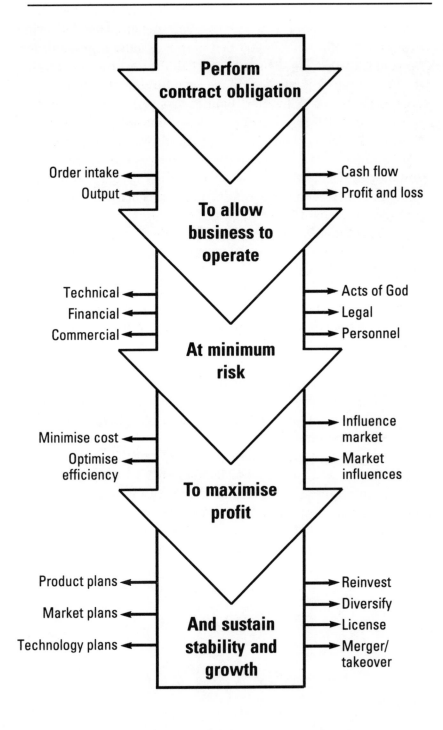

4

The contract

① Introduction

Each country has its own system of law and within each system are many divisions and subdivisions. In England the law divides into two:

Public law is concerned with the constitution and functions of the many different kinds of governmental organisations, including local authorities, and their legal relationship with the citizen and each other. Public law is also concerned with crime, which involves the state's relationship with, and power of control over, the individual.

Private law is concerned with the legal relationships of ordinary people in everyday transactions. It is also concerned with the legal position of companies. Private law includes contract and commercial law, the law of tort, law relating to family matters and the law of property.

There is also a division into criminal and civil law. Criminal law is concerned with legal rules which provide that certain forms of conduct will attract punishment by the state. Civil law embraces the whole of private law and all divisions of public law except criminal law. Fig 4.1 indicates the relationships.

The law in other countries varies dramatically in construction and application compared with our own. Only English law is considered here.

Figure 4.1: Structure of the law

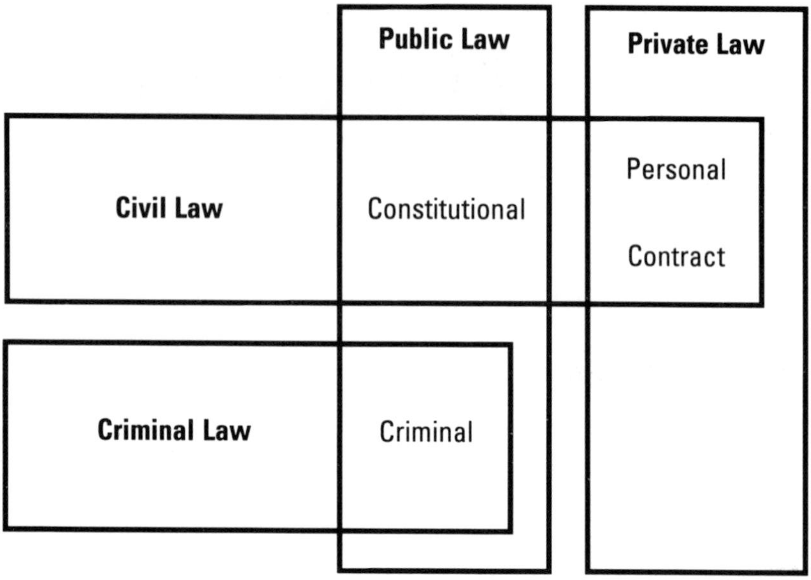

	Public Law	Private Law
Civil Law	Constitutional	Personal Contract
Criminal Law	Criminal	

Contract law, as well as falling into the categories of privatelaw and civil law, is one of the divisions of the system known as common law. This system originated in ancient customs which were clarified, extended and universalised by judges over the years. This process really began over 800 years ago. It is through this process that the concept of legal precedent came about, whereby a judge would simply rely upon a decision in a previous case to settle a new problem.

Although derived from common law, a contract and the contracting parties will be subject to various statutes (Acts of Parliament); for example, the Sale of Goods Act (1979) and the Unfair Contract Terms Act (1982).

International treaties and conventions and membership of the EEC can also affect the ways in which contracts may be created and discharged,as can the enormous raft of governmental regulations and procedures. The influence of the EEC is considered in Section 8 of this chapter.

② The Law of Contract

2.1. Types

Contracts can be simple or speciality. Speciality contracts are also known as contracts under seal or deeds. This type of contract accounts for a very small proportion of business and is not considered further. Similarly, other types of contractually binding relationships (for example, licences, agencies, consortia) are not dealt with in detail. Attention will be given to the type of contract that accounts for the majority of our concern – contracts for the supply of goods (being an example of a 'simple contract').

2.2. Definition

In the simplest terms, a contract is a mutual exchange of promises. The seller promises to supply goods or services and the buyer promises to pay. If certain criteria (explained in paragraphs 2.3 below) are satisfied, this exchange of promises can be enforced in a court of law. That is to say, the court will recognise that both people actually intended the promise to be *binding*. For this reason most organisations permit only certain of their employees to enter into contractual commitments on behalf of the organisation.

2.3. Key elements

The key elements that a contract must have in order to be valid and enforceable are:

a) An offer and an acceptance.

b) An intention by both parties to the contract to create legal relations.

c) Consideration.

d) The parties must have the capacity to contract.

e) The contract must be legal and possible.

2.3.1. Offer and acceptance

On the face of it, offer and acceptance is simple and straightforward. *A* offers to supply 10 widgets to *B* for £5 each. *B* accepts and a contract is created. However, if *A* advertises widgets at £5 each this is not an offer to sell but an invitation to treat. This would mean that *B* would have to offer to buy at £5 each and *A*'s acceptance would create the contract. It is thus important to be certain when an offer is actually being made. For example, the current practice in buying supermarket vegetables is for the shopper to make his selection and then at the checkout the cashier will say the price. This is the point at which an offer is made.

Acceptance also can have its complications. For the acceptance to create a contract it must be given without qualifications or conditions since to do so creates a counter-offer which itself must be accepted before a contract can come into being. Quite commonly it is the practice for a company to say to the customer 'we accept the contract subject to the following …'. Strictly speaking this is a counter-offer and no contract is made until the counter-offer has been accepted without qualification. As a matter of custom and practice the two parties to the 'contract' may each proceed with the business of the contract – the one to supply goods, the other to make payments – and a court may decide that a contract did indeed exist (this would be known as 'contract by performance'). The only question to be decided is whether or not *both* parties intended the qualifications given in the initial response to apply. Again, this may depend upon the actions of the parties.

For example, if the statement was to the effect that 'we accept the contract but will deliver blue widgets instead of green', and the customer, without having formally confirmed it, accepts deliveries of blue widgets, then clearly the qualification was mutually accepted. A factor to be taken into account is where the matter rested as far as correspondence was concerned. Where a matter has been debated without full resolution, whichever party had the final say in correspondence may well have the advantage.

2.3.2. Intention to create legal relations

As has already been said, a contract can be enforced by a court if the parties intended their promises to be binding. As a natural consequence of this, the court will provide remedies for the breaking of binding promises – known as breach of contract. In some circumstances it may be that promises made were intended to be kept but no one really expected or wanted a legal remedy for a broken promise – perhaps a cancelled invitation to dinner. In such cases a reasonable man (a standard if somewhat subjective test) would say that there could have been no intention to create legal relations and thus no contract was made.

In some ways the point being examined here is the question of people *by their actions* indicating or not indicating an intention to create legal relations. The whole purpose to professional life is the pursuit of business, the creating of contracts. It might be said therefore that there should be no doubt that the intention is to create legal relations. However, it is frequently just the opposite. Marketing, sales, engineering and projects people will regularly discuss possible transactions with potential customers and suppliers with no intention of creating legal relations. It is vital therefore that in such matters the purposes of the discussions are clear to all so that relations are not accordingly established. Quite often the contracts man may fire off a letter confirming the discussions but stating that 'the discussions and the letter do not constitute an order or a commitment to place an order with you'. This belt and braces disclaimer points out that not only is no order created but also that no intention to place an order may be construed from the actions and discussions.

Of importance to the individual in this context is that in, say, attending meetings he is representing his company in an official capacity and by his actions he may inadvertently commit the company. Although he may not have authority to make that commitment it may nevertheless stand unless it can be shown that the other side knew him not to have such authority.

2.3.3. Consideration

Consideration is the legal word for the money that is paid for the supply of goods. In fact, money is only one example of consideration, which has classically been defined as 'some right, interest, profit or benefit accruing to one party, or some forbearance, detriment, loss or responsibility given, suffered or undertaken by the other'.

Although somewhat unwieldy, this definition serves at least to convey the principle that the consideration must be of some value to the recipient. It is important to note also that the law does not exist to repair bad bargains. So the courts will examine the question of whether the consideration is sufficient to be valuable rather than of adequate value for the goods being supplied. The other major feature of consideration is that it must not be illegal! Generally speaking, the question of validity of consideration does not arise in most types of business transaction.

2.3.4. Capacity

The capacity to contract is fundamental. All adult citizens have the capacity to contract, although there are exceptions. For example, there are circumstances in which contracts made by aliens, persons suffering from mental disorder or drunkards are void. It is important to distinguish between capacity and authority to contract. The former is concerned with the legality, the latter is concerned with permission. In business an individual may have the legal capacity to contract but not the authority of his employer.

2.3.5. Legal and possible

The contract must not be illegal; for example, a contract to carry out a crime would not be a contract. Also the contract must be capable of performance. There could be no contract to supply a perpetual-motion machine for example.

2.4. Written v. oral

Simple contracts of the type being discussed do *not* have to be made in writing. As individuals, the majority of contracts we make are oral – whether it be buying a newspaper or purchasing a meal.

Generally, the only contracts that are required to be made in writing are those for the sale of: a house, flat, etc; long-term leases; those that guarantee another person's debt; hire purchase or consumer credit arrangements.

Companies as a matter of professional necessity adopt a practice of committing all contracts to writing. This is for several good reasons, including:

a) Unlike the typical consumer purchase, the subject-matter and rights and obligation of the parties may be extensive in description and definition. This naturally demands commitment to paper.

b) It is vital that both parties are clear and share the same understanding of the contract.

c) As individuals come and go it is important that their successors can establish clearly what is involved.

d) A written contract is a sound baseline for changes in requirements, rights and obligations which may arise and become contract amendments.

e) In the event of a dispute during or after the completion of the contract the court will be better able to reach a decision based on written evidence.

f) Where many functions within the company will exchange correspondence with their opposite numbers (on project, engineering or marketing networks for example) it is important to know which bits of paper actually constitute the contract.

Although written contracts may be the preferred approach, most companies will both place and accept oral contracts where the urgency of the situation demands. The aim will of course be to reduce these oral contracts to writing as quickly as possible. Nevertheless, to be valid, even oral contracts must satisfy the basic legal requirements described above.

In today's technological age even computers have a role to play in the actual placing of contracts. One party may agree to accept computer-generated orders from another party. Once again, though, the fundamental principles apply, i.e. the computer order in practice would constitute an offer which the other party would have to accept before the contract is created.

2.5. Terms and conditions

'Terms and conditions' is an often-heard expression which tends to be dismissed as the 'contractual bit'. The expression itself is to some extent misleading.

The terms of the contract are all of those things which describe all of the rights and obligations of the parties. The terms address, indeed constitute, the entire description of the contract.

Terms are either express or implied, Express terms are those which the parties themselves have established and agreed to. Implied terms are those terms which either:

a) a court will decide may be read into the contract based on what the parties must have intended; or

b) arise from a statute which establishes terms that apply whether or not the parties intended (or even thought) that they should.

The best example of this is the famous Sale of Goods Act, which, amongst other things, implies terms of fitness for purpose and merchantable quality.

The terms of the contract are also subdivided into conditions and warranties. Not all of the obligations created by a contract are of equal importance and this is recognised by the law, which has applied a special terminology to contractual terms to distinguish the vital or fundamental obligations from the less vital. The word 'condition' applies to the former and warranty to the later. 'Warranty' in this sense should not be confused with the common usage relating to a supplier's guarantee. This is considered in more detail later. Fuller definitions of condition and warranty might be as follows:

'A condition is a vital term which goes to the root of the contract. It is an obligation which is so essential that its non-performance may be considered by the other party as failure to perform the contract at all.'

'A warranty is subsidiary to the main purpose and is not so vital that a failure, to perform it goes to the substance of the contract.'

To put this into the context of a contract for the supply of goods the conditions would be, on the one hand, the supplier to supply the correct goods and, on the other, the buyer to make payments. A warranty would perhaps be for the supplier to provide regular progress reports.

In practice, many purchasers will refer simply to the 'conditions of contract', implying that all the terms are fundamental. This would not necessarily be upheld by a court. A crucial difference between conditions and warranties is the remedies that can apply in the event of a breach of contract.

2.6. Breach of contract

The law provides several remedies for breach of contract that depend upon the circumstances of the breach and the nature and extent of the consequences suffered by the injured party. In the case of breach of a contract condition, the injured party is entitled to end the contract and pursue the defaulting party for damages. Breach of a contract warranty entitles the injured party to sue for damages only, i.e. if it is not a fundamental feature of the contract the injured party has to continue with the contract.

The purpose here is not to provide an exhaustive analysis of breach and remedies, rather it is to put across one of the principles upon which contract law is founded. In practice, the contract terms themselves may provide remedies where one party is not fulfilling his obligations. Commonly, this will include the right to delay payment or recover payments already made.

2.7. Damages (compensation)

Damages are the compensation that a court may award for injury, loss or damage arising as a result of someone else's action or failure to act (including, for example, breach of contract). The court would have to be satisfied that the injury, loss or damage was directly caused by the action or inaction. Injury, loss or damage caused indirectly by the action or inaction may not lead to the award of damages unless the contract provides that one or both parties accept liability for indirect injury, loss or damage caused to the other by their action or inaction.

③ Application of the law

The aim of the foregoing has been to describe the framework of the law within which contracts must be made and to briefly touch on what the law provides if things go wrong. However, it would be wrong to form the impression that buyers and sellers are always dashing off to court to settle questions of interpretation and performance, or non-performance, as the case may be.

In the UK, contracting parties tend to the view that both sides have lost if the problem ends up in court, and for the most part every effort will be made to settle the difficulty by sensible negotiation. At the end of the day the buyer wants his goods and the seller wants his money, neither side believing that his aim will necessarily be achieved where there is the potential cost and delay of legal action to contend with.

Furthermore, a supplier will be most unwilling to cause upset to his customer by taking him to court (or even threatening or suggesting this would happen) unless there really is no other option.

Indeed the contract may set out extensive procedures for handling disputes, whether of a technical or commercial nature. This may include a mechanism for negotiation which, if unsuccessful within a specified period of time, will allow the two sides to elevate the problem within their own organisations until a level is found at which a settlement can be reached. In some cases one side (usually the buyer) may reserve the right to have a final and conclusive say on a particular aspect, interpretation of technical specification for example.

In between the processes of full legal action through the courts and a disputes procedure written into the contract lies the process of arbitration. The contract itself will usually provide for settlement by arbitration in line with the Arbitration Act 1950.

The sensible and pragmatic use of the law followed in this country contrasts dramatically with the practices in other countries, where use of litigation in commerce is commonplace.

(4) Instructions to proceed and intentions to proceed

The text so far has dealt with some aspects of contract law, the formation of a contract, and remedies for breach. All of this presupposes that there is time to put a full contract in place. In business it is not infrequently the case that, because of some urgency, or indeed for other reasons as well, the customer would like the supplier to proceed in advance of establishing a full contract. For reasons of his own, the supplier may be prepared to go ahead where, for example, a later start would give him difficulty with availability of resources. the infamous acronym ITP is then bandied around. These letters usually have two interpretations.

Intention to Proceed (also known as a letter of intent): this form of ITP means exactly what it says – the buyer has an intention to purchase. The value of this to the supplier is only that it gives him a clear idea of what the buyer might want and perhaps gives him more confidence that a sale is close. An intention to proceed does not create a contract and the supplier, if he proceeds with work, does so at his own risk.

The expression 'working at his own risk' conveys that the risk is not recovering the cost of the work done. In specialist areas this work may be completely nugatory in so far as it may not even be of any use in meeting the requirements of other customers. On top of this the actual doing of the work will have diverted effort from real contracts where there are profits to be made. It cannot be stressed heavily enough that there are real dangers in doing what it is thought the customer wants rather than that for which he has actually contracted.

Instruction to Proceed. This ITP is intended to create a binding relationship. Provided that the ITP satisfies all the fundamentals of a contract and provided that it is accepted, a contract is created. This will be used where time does not permit the buyer to set out his full requirements but nevertheless sufficient of the contract conditions can be specified.

It is most important that where an ITP has been issued it is established whether it is an instruction or an intention only. Wherever possible an instruction to proceed will be given in writing, but, as with oral contracts, the aim will be to formulate the full contract documents as quickly as possible.

⑤ Essential elements in practice

5.1. Introduction

Having dealt with the legal principles on which a contract is formulated it is important to examine the elements that are essential in practice, particularly in so far as the project managers and engineers responsible for performing the contract are concerned. This is the what, when, where, how and what else of the contract. The degree of certainty that can be attached to each of these influences the risk involved in, and cost of, doing the job.

5.2. The 'what' (definition and specification)

It is vital that the goods are defined as absolutely as possible. In the simple case the 'what' will be part number, description and quantity, but it is also important to specify other things to be supplied – handbooks and spares for example.

Where the 'goods' are particularly complicated – where design and development or complex systems are involved – engineering standards, specifications, acceptance and handover arrangements are needed. Deliverable data and information in format and content must be given. Project control, monitoring and other services required must be specified.

5.3. The 'when' (timescale and timing of contract performance)

The contract should specify when and at what rate things should be delivered. The buyer will attempt to have delivery dates written into the contract. The supplier will attempt to minimise the firmness of his commitment either by referring to so many weeks from receipt of order or date of contract or by avoiding a specific delivery promise altogether.

The buyer may specify that time of delivery is of fundamental importance to him. This is known as 'time is of the essence of the contract'. The buyer is clearly saying that making delivery at a particular time is a condition of the contract, and as has already been said, this entitles the buyer to cancel the contract and sue for damages. The implication of a time-is-of-the-essence clause is that the buyer will suffer damage if contract performance is late. The seller will usually seek protection against this in two ways. Firstly he will attempt to negotiate a 'liquidated damages' clause. The aim is to agree in advance with the buyer the likely extent in financial terms of the consequence of late delivery and then to include in the contract a mechanism for liquidating this sum. A simple example of this would be as follows:

'For every week that the goods are in delay the seller shall pay to the buyer the sum of £X subject to an overall limit of £Y.'

It is common practice for a buyer to actually include a liquidated damages clause though there may not be a 'time is of the essence' statement. The objective here, fairly obviously, is to create an incentive on the seller to deliver quickly. Strictly speaking, this is to create a penalty contract which is not enforceable in English law. On many occasions the seller may simply, but reluctantly, accept this rather than to seek to have the legality tested in court.

The second source in protection that a seller will seek is a *force majeure* or 'excusable delays' clause. This will permit the seller an extension of time in which to perform the contract, and consequently a suspension of the application of the liquidated damages provision, if he is delayed by, for example, Acts of God. It should not be thought, however, that *force majeure* is an automatic 'get out' as:

a) there will be a duty on the supplier to mitigate the effect of the delay;

b) in any event he may not be excused the delay if his insurance covers the effect of the incident involved;

c) many categories of delay – failure of a subcontractor for example – will be considered to be within the control of the supplier and therefore outside of the *force majeure* protection.

5.4. The 'where' (destination)

Where are the goods required? The contract will give consignment instructions. This does not necessarily require the supplier to deliver the goods to the consignee, only that he should address packages, etc; accordingly. This is considered in more detail in Chapter 6.

5.5. The 'how' (method of physical delivery)

How will goods be delivered? By road, rail, sea or air? Package and preserved how? Who will deliver, the supplier or his carrier, or will the buyer collect? These issues will be addressed in the contract.

5.6. The 'what else' (dependencies)

What other things does performance of the contract rely upon? Does the buyer have to provide goods (perhaps for embodiment or testing purposes), services, facilities, data or information in order for the supplier to discharge his responsibilities? Again, these aspects will need to be included in the contract.

5.7. The risk

The purpose of going over what may appear to be simple and obvious points in this 'what, when, where, how, what else' analysis is to draw attention to the fact that all the things mentioned have a cost impact.

Where there is uncertainty over the obligations of the contract there is uncertainty over the cost and consequently a risk to the commercial successes of the venture. This risk applies to both the buyer and the seller. The buyer must ask himself 'can I be sure I will get what I want when I want it', the seller must ask himself 'can I be sure I can complete the contract quickly and efficiently in the manner that I have planned'. Where there is any substantial degree of technical complexity involved in the contract the engineer clearly has a key role to play in ensuring the maximum certainty and therefore the minimum risk.

5.8. Price and payment

Last but by no means least are the essential elements of price and payment. Both these topics are later dealt with in more detail. At this stage, suffice it to say that the contract must specify the price, or where it is not possible to specify the price at the time the contract is made, then the mechanism by which a price will be agreed must be included. Similarly it is important to indicate how and when payments will be made.

5.9. Contract number

A final point to note is the relevance of the contract number. This is not actually a legal requirement but nevertheless an essential element in so far as good business practice is concerned. Certainly a lot of companies will not usually accept an instruction to proceed unless it includes a contract or order number. The inclusion of the number will at least give the recipient the confidence that action is probably in hand to produce the full documents. More importantly, should work proceed under the ITP to the point when some payment is due, it is unlikely that the buyer's purchasing/payment system will recognise and accept an invoice unless it is possible to refer to a valid contract or purchase order number.

5.10. Contract amendments and changes

Before moving on to discuss contract layout it is worth mentioning the function and meaning of amendments and changes to contract.

Functionally, an amendment to contract has identical status to a contract itself. That is, before it can become a binding part of the contract there must be offer and acceptance, valuable consideration and intention to create legal relations.

All the rules apply but in practice some aspects may be handled differently. For example, the contract amendment, being a contract in its own right, should identify all the terms that are to apply. However, in a large number of cases the parties will wish the terms of the existing contract to apply, and rather than simply repeat them all the wording of the contract amendment will close with a phrase such as 'all other terms and conditions remain unchanged'. As far as consideration is concerned, the wording of the contract may permit amendments to occur with the extra consideration to be arrived at later, but in a prescribed manner.

For practical purposes the crucial point regarding amendments is that they must be accepted before they can become binding. It is important to be certain whether an amendment has been accepted or not. Confusion can be caused if the customer has the practice of distributing copies of amendments on the day of issue of the amendment (at this stage it is strictly speaking a *proposed* amendment only) rather than on the day of acceptance.

Contract changes are a slightly different device to the contract amendment. Where the parties can predict that there will be, say, many technical changes during the life of the contract they may wish to avoid the full formality of contract amendments as a means of control. Instead they may choose to set down in the contract a mechanism for the changes to be controlled and authorised. Provided the mechanism is clearly set down, and provided that also set down is how and when prices and payments for the changes will be agreed and made, all is well.

5.11. Layout of a contract

The physical size of the contract documents almost always vary in proportion to the complexity and value of the task. The purchase of a standard product at a standard price may involve nothing more than the buyer's standard purchase order, comprising a single sheet of paper, the front side describing the item, delivery and price, the reverse side containing pre-printed conditions of contract. the supplier may not necessarily agree with all these conditions, but if, as a matter of practice, they have never given him a problem he may choose not to make an issue of it and risk losing a customer.

The layout of a major contract does not have to follow a pre-set form but the following is a typical example.

Section 1: Priced schedule of requirements.

Section 2: Statement of work.

Section 3: Specifications.

Section 4: Contract conditions.

Section 5: Payment plan.

Section 6: Definition of deliverable articles.

Section 7: Definition of deliverable data.

Section 8: Quality plan.

All or part of each or all of these section will have been the subject of detailed negotiation which may take weeks, months or years to agree.

Another factor is the extent to which documents forming part of the contract are physically included or included by reference. Where parties regularly do business with one another they will aim to reduce as many of the regular features as possible to standard forms which can be incorporated by reference. This practice itself and standard interpretations that inevitably arise from the use of standard forms helps to create 'custom and practice'. This allows both buyer and seller some degree of comfort and confidence in the proceedings.

6 Contract conditions

6.1. Standard conditions

Where one or more parties regularly do business with one another there is advantage to all concerned in the use of standard conditions; that is, a set of conditions that buyer and seller will negotiate and agree as the basis on which most, if not all, of their transactions will take place. This saves considerable time in the drafting of contract documents and indeed in the physical bulk of the documents, as the standard conditions can be included by reference. Not only that but the contracting parties will understand and be confident in the custom and practice that has grown up around the use of standard conditions. Standard conditions fall into three categories:

a) Those adopted by government purchasing departments.

b) Those adopted by industry or a sector of industry as a basis for commercial transactions.

c) Those unique to a particular firm, either as standard conditions of sale or as standard conditions of purchase.

In practice, the use of these categories of standards may overlap. For example, many HMG standard conditions include specific obligations to flow down equivalent conditions to subcontractors. Thus a subcontractor dealing with an HMG contractor may not be in a true 'commercial' arrangement, subject to one of the industry sets of conditions, as he is performing HMG work and the appropriate conditions must apply. The extent and level to which this flow down must be taken is dependent on several things. For example, the flow down of conditions relating to the agreement of prices and earned profit is set strictly by value. On the other hand, conditions relating to design rights and patents are usually to be included in all subcontracts where design or development work is undertaken. Other issues are matters of commercial judgement. For example, if HMG requires warranty then it is prudent for the main contractor to secure warranties from his suppliers, even if there is no mandatory flow down of warranty.

HMG standard conditions are negotiated and agreed at the macro level betwen HMG and industry, usually in the person of the CBI. The CBI, with its many committees and subcommittees and other trade associations feeding into it, clearly has a difficult task in soliciting and co-ordinating opinions and views from the very many member firms and organisations. It is not only a question of co-ordinating comments but of fairly and effectively representing the needs of an enormous variety of companies in terms of size, location and technology group. Similarly, HMG has several major purchasing departments, ranging from the DSS to the MOD.

The concerns connected with purchasing for hospitals in Birmingham are somewhat different to those relating to airborne radar flying at 100,000 feet. Necessarily therefore the negotiation and agreement of standard HMG conditions can be long-drawn-out process. Changes needed through the evolution of technology, economic changes, government policy and industry's needs also take some time to come to fruition. HMG will publish the new standard regardless and then negotiation of an acceptable compromise becomes a matter for individual contract discussions. In such circumstances HMG sometimes hopes that the new standard will become established simply by its acceptance in a significant number of contracts.

Commercial standard conditions are generally published as sets of conditions sponsored by trade associations in particular industry sectors, principally these are:

The Institute of Electrical Engineers.

The Institute of Mechanical Engineers.

The Institute of Civil Engineers.

The Federation of British Electrotechnical and Allied Manufacturer's Associations (BEAMA) publishes various of these and their own sets of standard conditions. A list of those currently available is included at Annexe A.

Standard conditions of sale or purchase specific to individual firms tend to be printed on a single side in very small print and appear on the reverse side of official paperwork. The front side of a purchase order might say 'subject to our standard conditions of purchase'. The supplier's invoice may have a statement to the effect: 'subject to our standard conditions of sale'. In the event of a dispute the court may have some judgement to exercise as to which conditions the two parties intended to apply to the contract. These naturally will vary between company and company, depending on the nature of the goods and the commercial strength of the company. For example, a firm supplying paper-clips where there are dozens of competitors may draft a fairly generous set of standard conditions of sale so as not to cause friction with his customers. On the other hand a monopoly, or virtual monopoly supplier may chose a tough set of conditions as he can trade on a take-it-or-leave-it basis.

One other form of standardisation of conditions is where the first one or two contracts between two parties are fully negotiated, but as subsequent contracts of a similar value and type are placed, the parties may agree to adopt the same set of conditions that applied previously.

Other organisations, such as the Institute of Purchasing and Supply, publish model conditions. Model conditions are really indended as drafting aids whereas true standard conditions are intended to be used as they stand without qualification or caveat.

6.2. Negotiated conditions

At the end of the day, contract conditions can only be reduced to standards where buyer and/or seller see no advantage to the time-consuming activity of negotiating conditions on each and every occasion that a transaction is contemplated. Also, some aspects are actually different every time a transaction is entered into and could thus never be reduced to a standard. For example, payment terms in industries where interim payments are permitted can vary on each contract and as a key commercial matter most companies will wish to negotiate these on a case-by-case basis. A key commercial skill with regard to negotiating contract conditions is the trade- off analysis between genuine risk, practical solutions and legal niceties. For example, a condition which permits the customer to cancel the contract for convenience might be perfectly acceptable legally if the wording is correct but commercially a disaster unless an adequate period of notice must be given. On the other hand a ludicrously brief period of notice might be acceptable if the risk of cancellation is highly remote.

The relative proportions of standard and negotiated conditions will depend on the value and nature of the work. Where the goods are low value off the shelf the buyer's or seller's own standard conditions may be entirely acceptable. For a major contract, especially those involving design and development, it could well be that any standard conditions proposed by one side can only be used as a place to start from in negotiation. That is to say, a condition which involves some risk to one party may not be worth arguing about in the purchase of catalogue components but could be a crucial factor in the negotiation of a brand new jet engine.

Where the basic legal requirements have been satisfied (offer/acceptance, consideration, etc) and a binding contract created, the parties may agree to proceed with the performances of the contract although some conditions remain to be negotiated and agreed in detail. In these circumstances one party may perceive advantage in delaying agreement, particularly if it is something the full impact of which was not earlier appreciated. For example, a requirement to adopt a new engineering or component standard could be costly if the supplier does not have the opportunity to recover the cost of changing, and thus he may delay agreement by drawing out negotiation of the contract condition to the point where it is too late to implement the change anyway. On the other hand, the buyer may delay agreement on the wording of the payment condition so as to delay the evil day of making payments. The experienced commercial negotiator will be alert to these tactics and will employ appropriate counter-measures.

Conclusion

This chapter has served to introduce the framework of the law in which contract law resides and the main principles of establishing a binding contract. Some of the essential elements in practice the what, when, where, how – have been explored. Having considered these issues, the contracting parties will give thought to what else should be built into the contract – whether by the use of standard and/or negotiated conditions – to protect their position, minimise risk and to put the maximum possible obligation, responsibility and liability on the other side. Although the various standard conditions, for example, appear to be many and varied, in practice there are very many common principles which are simply worded differently.

(8) The impact of 1992

Although much public attention is focused on the harmonisation of the European Community planned to occur in 1992, the conduct of business by UK companies has for a long time been subject to Community Law. The European Communities Act 1972 incorporated Community Law into the English system of law. The effect of this is that English courts must give precedence to European Law where there is either no applicable English Law or where there is a conflict between the two.

There are two types of EC requirements:

1) A regulation: this is automatically binding upon UK and does not require UK legislation for it to be effective.

2) A directive: in this the intended result is binding, but national legislation is necessary to cover its implementation.

In both cases Parliament seeks to enact legislation to ensure that the courts have a reference from which to make judgments, albeit that these must be construed in accordance with the European Law.

Annexe A: BEAMA legal department price guide:

BEAMA Publications: Revised April 1986 – unless otherwise stated.

The following conditions are sold in minimum quantities of 100:	**£**
Conditions of Sale A:	
For Machinery and Equipment (excluding Erection), UK	7.00
Conditions of Sale AE:	
For Machinery and Equipment (excluding Erection) Export: FOB, FOR, FOT	7.00
Conditions of Sale AEC:	
For Machinery and Equipment (excluding Erection) Export: CIF and F & F	7.00
Conditions of Sale B:	
For Machinery and Equipment (including Erection), UK	9.00
Conditions of Sale BE:	
For Machinery and Equipment (including Erection) Export: FCB	9.00
Conditions of Sale C:	
For Electronic Equipment Including Installation, UK	9.00
Conditions of Contract:	
For commissioning Electronic Equipment, UK	7.00
Conditions of Sale E:	
For Erection of Electrical Plant and Machinery (Home or Export)	9.00
Conditions of Sale R:	
For Repair of Machinery and Equipment. UK	7.00
Conditions of Sale RE:	
For Repair of Machinery and Equipment, Export: FOB	7.00
Conditions of Sale SA:	
For Stock and Catalogue Articles, UK	7.00
Conditions of Sale SAE:	
For Stock and Catalogue Articles. Export: FOB	7.00

The following conditions are sold singly:

Conditions of Sales RC:	
For Reconstruction, Modification or Repair of Plant and Equipment, UK	4.00
Conditions of Contract:	
For Systems Incorporating Electronic Equipment (Home or Export)	3.00
Conditions of Contract:	
S & E For Supply, Erection and Commissioning of Electrical Plant andd Machinery	2.00
The BEAMA 'White Book':	
Comprising one copy of each of the above conditions, bound in book form	10.00
EB/BEAMA 1979 (A):	
Conditions of Contract for Plant including Erection	5.00
EB/BEAMA 1979 (B):	
Conditions of Contract for Plant without Erection	5.00
EB/BEAMA 1984 (RC):	
Conditions of Contract for Works of Reconstruction, Modification or Reapir of Plant and Equipment, Involving Work on Site	5.00
Contracts for Acquisition and Utilisation of Computer Software and Monitoring Systems (1982 edition)	20.00
'Export Contracts':	
Course Notes from a recent BEAMA Seminar	15.00

IEE/I.MECH.E./Model conditions of contract:

Model Form A: Home Contracts with Erection	4.50
Model Form B1: Supply of Plant and Machinery (Export)	4.50
Model Form B2: Delivery (with Supervision of Erection) Export FOB, CIF or FOR	4.50
Model Form B3: Export Contracts (with Delivery to and Erection on Site)	4.50
Model Form C: Supply of Electrical and Mechanical Goods (Home Contracts – without Erection).	4.50
National Water Council Form G: Water Authority Plant Contracts Standard Amendments and Additions to Model Form A (Home Contracts with Erection).	3.50

Please note: Orders Will Only be Accepted When Accompanied by Your Remittance.
Cheques Should Be Made Payable to 'BEAMA Ltd' and forwarded to the address detailed below:

The Legal Department, BEAMA Ltd, Leicester House, 8 Leicester Street, London WC2H 7BN.

Annexe B: BEAMA legal department price guide: Orgalime Publications

£

General Conditions for the import and export of semi-processed goods and components for incorporation in other goods	1.50
Model form of exclusive agency contract with a foreign company or individual	5.00
Model form of exclusive agreement with distributors abroad	5.00
Model form of maintenance contract	5.00
Model form of processing contract	5.00
Model form of licence agreement for the manufacture of an unpatented product	5.00
Conditions for the provision of technical personnel abroad	5.00
Practical guide for preparing a know-how contract	5.00
Drawings and technical documents – ownership and protection against improper use	8.50
Progress terms of payment in the Electrical Industry	8.50
The guarantee or defects liability period	8.50
Contract price adjustment	8.50
Guide for drawing up an international consortium agreement	10.50
Guide for drawing up an international research and development contract	10.50
Security for payment in sales with deferred payment	10.50
The liability of the seller for latent defects	10.50
Validity of the EEC conditions in Western European countries	10.50
Liability of the sub-contractor under ECE conditions towards the customer of the main contractor	10.50
Delay in delivery, erection and payment	10.50

Postage and packaging are included in prices quoted
Note: remittance must accompany all orders

5

Cost and price

① **Introduction**

The buyer wants to buy and a seller wants to sell. Between them they must agree a price as an essential element in the formulation of a binding contract. In the world at large experience shows that prices can be set as follows:

a) By advertisement. Displayed with the goods on a shelf or in a showroom or recorded in a catalogue or price list.

b) By competition. This falls into two divisions. Who will pay the most – as in an auction – where a monopoly supplier has several potential customers. Who will charge the least, where a readily understood customer has several potential suppliers.

c) By negotiation. On the simple level this could be unsophisticated 'horse-trading'. On the complex level regard will be paid to the suppliers cost and the degree of reward to which he should be entitled.

Another way of looking at this is to consider that there are two questions only:

a) What can the buyer afford to pay?

b) At what price can the supplier afford to sell such that he can sustain and develop his business?

Whether the agreed price is established by advertisement, competition, negotiation or combinations of these three, and on the assumption that the price is affordable by the customer and offers the supplier a reasonable return, then the question is: from where did the price come in the first place? The price to be displayed in advertisement or offered as an opening position in a negotiated arrangement has to be calculated. This chapter looks at the construction of a price and the different types of price that can be formulated.

② Terminology

For the purposes of this discussion these terms will be defined as follows:

'Cost', is the sum of money needed to do the job. 'Price' is the some of money paid by the customer. The difference between the two, if positive, is profit, if negative, is loss.

In pure accountancy the measurement of profit is more complicated than this and there are many different ways of defining it. This is of particular importance to companies where the nature of the work, perhaps a high turnover of standard products at standard prices, means that profit is only measured by accountants and by the company as a whole. On a major project where the project manager will be charged with responsibility for making the contract profitable, a measure of price against cost is most important.

Having defined cost, profit and price it is necessary to define the difference between estimates and quotations:

Estimates for all intents and purposes are guesses. This is something of an exageration because a guess is a figure pulled out of the air and an estimate is an attempt to indicate a figure that is reasonably close (perhaps ±15%) to what the real figure will be. The main point is that an estimate is usually not intended to be a firm offer which the other side could accept to create a binding contract.

Quotations are intended to be an offer which the customer can absolutely rely upon and, if he wishes, accept, to create a contract. The seller will usually, in his quotation, specify a validity of time commencing with the date of quotation within which the quotation may be accepted to create the contract. After the expiry of the period the quotation automatically lapses and is no longer capable of acceptance. The seller may also reserve the right to withdraw or amend the quotation during its validity period but only prior to acceptance by the customer.

The foregoing is the general rule. Custom and practice in particular industries or businesses may reverse these definitions. The estimate, sometimes known as a budgetary estimate, would be seen as an order-of-magnitude guide to the customer to let him work out if he can afford it at all, and if so to permit him to allocate funds in his budgeting. Preparation of an estimate is less time- and effort-consuming than a quotation. Therefore before committing to the preparation of a quotation it is worth finding out if the customer actually wants a quotation, i.e. if acceptable an order will be placed.

Contracts can be classified by the price type, i.e.

a) Firm price: in which the price is not variable for any reason other than a change to the work content or duration.

b) Fixed price: in which the final value is set (i.e. fixed) by reference to some variable factor such as an exchange rate or cost-escalation index.

c) Cost plus: in which the seller is reimbursed his actual costs plus a fixed or percentage fee.

d) Incentive: in which the price is ultimately determined by the supplier's success in meeting or beating established targets of cost or time-scale or technical features.

Readers should not rely absolutely on these definitions. In particular, the definitions of fixed price and firm price are transposed in different trade practices and are sometimes combined as fixed firm prices or firm fixed prices just to cause total confusion. It is important to discover which convention applies.

③ Elements of cost

In selling products to customers at whatever the prices may be, the supplier must aim at the very least to recover all of the costs in manufacturing the product and making it available to the customer. Consider what these costs might comprise:

Factory set-up
   Purchase of the land
   Solicitors' fees
   Survey fees
   Government fees (stamp duty, etc.)
   Bank costs

      Building costs
      Purchase of plant and equipment
      Delivery and installation of plant and equipment
      Recruitment costs (advertising, interviews, etc.)
    Factory operation
      Heating
      Lighting
      Rates and rent
      Repair and maintenance
      Interest charges
      Taxation
      Advertising and marketing expenses
      Depreciation of buildings and plant
      Provision of delivery or other vehicles
    Salaries and wages
      Research and development
      Manufacturing
      Machine operators
      Production control
      Quality assurance
      Project management
      Sales and marketing
      Finance and accounting
      Personnel
      Legal and contracts/commercial
      Goods inwards
      Despatch
      Security
    Materials
      Components
      Raw materials
      Subcontract work
      Purchase of special services
      Consumables – stationery, etc.
      Packaging and postage

    In the simple case of a single-product firm with a steady turnover the unit price might be calculated as the sum of the annual costs of the above less any government/local-authority grants, rebates and allowances plus something for profit divided by the total number of units made in a year. Clearly, where the turnover is not steady, where there are many products sold at varying quantities, then this highly simplistic calculation will not work. From this situation derives the concept of direct and indirect costs. Indirect costs are usually referred to as overheads. Direct costs are those costs which can be identified as being directly attributable to specific products. Thus machine-operator wages and development-engineer salaries would usually be direct. Personnel salaries which are not obviously attributable to individual products would be overheads. Purchase of components for the type-16 widget would be direct. Stationery expenditure would be overheads. Factory set-up and operating costs are generally overhead charges. The price for individual products can thus be built up from the direct costs attributable to each product plus a share of the overhead costs. The share may be determined on a simple mathematical basis, perhaps by recovering the overheads against each product *pro rata* against the respective volumes of manufacturing. On the other hand, if some products can be sold at a higher profit margin than others then perhaps they will absorb a higher proportion of the overhead burden than less-profitable products.

④ Estimating

Prices can be built up from direct costs and overheads. However, prices may have to be established before the product has been designed or manufactured and here the skill is in estimating the costs. A distinction should be drawn between an estimate given to the customer for his budgetary purposes and the internal task of estimating from which prices will be calculated for the purposes of providing the customer with a quotation.

The accuracy of estimating is an absolutely key element in the commercial success of the business. If the estimates are too high (that is, the cost of doing the job actually turns out to be less than the estimate) the amount of profit will be greater, assuming of course that the overestimate did not so unnecessarily inflate the price that the product did not sell in the first place. On the other hand, if the estimates are too low the product may make a loss in its selling price. Then again, if the effect of the artificially low price dramatically boosts the volume of sales then the opposite can happen, i.e. whilst the direct costs of manufacturing will vary approximately in direct proportion to the numbers sold, the overhead costs will tend not to be volume-dependent, thus an increase in sales can over-recover the overhead costs which might affect, or more than offset, the effect of underestimating the direct costs of manufacture.

The following are illustrative examples:

1) Overestimate (predicted sales achieved)

		£
1)	Total direct costs (estimated)	10,000
2)	Total overheads	5,000
3)	Total profit forecast	1,500
4)	Total price	16,500
5)	Predicted volume of sales = 1,000	
6)	Selling price each £16.50	
7)	Actual sales = 1,000 (as predicted)	
8)	Total direct costs (actual)	5,000
	(i.e. 50% overestimate)	
9)	Total overheads	5,000
10)	Total cost	10,000
11)	Total receipts = 1,000 × £16.50 =	16,500
12)	Profit earned = £16,500 − £10,000 =	6,500

2) Overestimate (predicted sales not achieved)

Figures as (1) above except sales only 500

		£
8)	Total direct costs (actual)	2,500
	(i.e. 50% overestimate)	
9)	Total overheads	5,000
10)	Total cost	7,500
11)	Total receipts = 500 × £16.50 =	8,250
12)	Profit earned = £8,250 − £7,500 =	725

3) Underestimate (predicted sales achieved)
Figures as (1) above

		£
8)	Total direct costs (actual) (i.e. 50% underestimate)	20,000
9)	Total overheads	5,000
10)	Total cost	25,000
11)	Total receipts = 1,000 × £16.50 =	16,500
12)	Profit earned = £16,500 − £25,000 = **a loss of**	8,500

4) Underestimate (predicted sales exceeded)
Figures at (1) above except sales = 2,000

		£
8)	Total direct costs (actual) (i.e. 50% underestimate)	40,000
9)	Total overheads	5,000
10)	Total cost	45,000
11)	Total receipts = 2,000 × £16.50 =	33,000
12)	Profit earned = £33,000 − £45,000 = **a loss of**	12,000

The aim is to illustrate some simple principles only. The presumptions made above may not hold true. For example, the direct costs may not be exactly directly proportional to volume. For small quantities there may be a cost penalty in the purchase of components or other materials in less-than-economic quantities. Similarly, overheads are to a certain extent volume-dependent. The greater the volume the greater the fuel consumption; more machine time means more electricity costs, etc.

5 Cost and time

The principle point being made is that accuracy of estimating is vital. So far we have considered this in terms of estimating, say, the total number or unit number of labour hours and again the total or unit cost of materials (together being the total or unit direct costs); but what of time? When will the labour actually be deployed, when will the materials be purchased? If the period of work is spread over months or years it is important to know the spread of direct costs – where labour costs in terms of wages and salaries will annually increase; where supplier quotations are only valid for a short period of time it is essential that the labour profile and material expenditure profile are known.

To develop a price for a two-year contract assuming an even spread of direct costs over both years could lead to a loss if the majority of direct costs fall in the second year at higher salary rates, etc.

⑥ Cost and performance

The foregoing discussion has centred on the importance of accuracy in estimating total or unit direct costs and the phasing thereof. In practice the price is agreed and will have been based on these estimates. Once the price is agreed there is usually little that can be done about inaccurate estimating. Once the contract is won at an agreed price the emphasis shifts from accuracy in estimating to efficiency in performance. If, through performance efficiencies, the direct costs fall then that saving is retained by the contractor as an extra profit. If the performance efficiency decreases the direct costs will rise and the profit margin can be eroded to the extent where a loss is made. In simple terms, the amount of profit is related to cost as illustrated in Fig. 5.1.

Efficiency or performance can be improved in many ways, e.g.

a) Decreasing the number of direct labour hours deployed.

b) Deploying direct labour effort as early as possible to take advantage of cheaper labour rates, etc.

c) Buying materials as early as possible and in as large quantities as possible.

Clearly there will always be a number of other factors to be taken into account. For example, employing direct labour as early as possible is a good idea only if there is other work for the labour force to do once the particular job has been finished. It may actually be preferable to suffer the cost penalty of direct labour being deployed late than to run the risk of laying people off or breaking up successful teams.

Figure 5.1

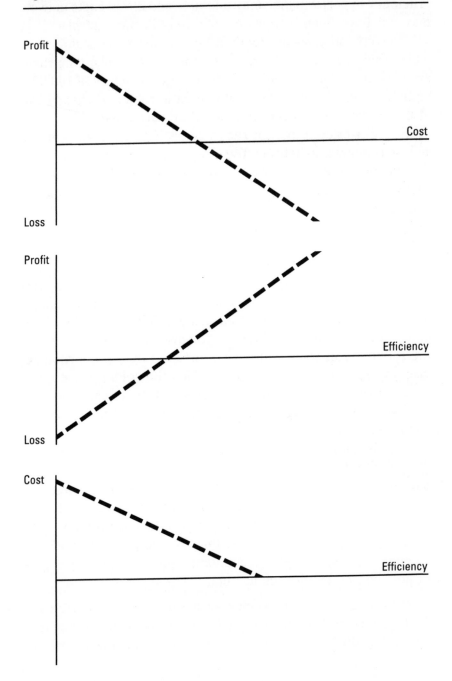

Overhead recovery

Direct costs are quite volatile. More work means more costs. Effort can be deployed early or late as circumstances dictate, efficiency of performance may fluctuate and direct costs may or may not have been estimated accurately in the first place.

Overhead costs are by their very nature more stable, and whilst this may assist to some extent in business planning they are of course more difficult to reduce. Once a factory of 100,000 sq.ft. has been set up and put into operation at an overhead cost of £2m a year, then a certain amount of work must be secured at the right price to permit recovery of that overhead. This is a close relationship between the direct cost of work put through the factory and the recovery of overheads. For example, securing contracts that will generate 2 million hours of direct work a year would permit a recovery of £1/hour. Sales can then be quoted on the basis of say £10/hour for direct labour plus £1/hour for overhead recovery, and this, let us presume, yields a selling price which is competitive in the market-place.

However, if things go poorly and in the succeeding year contracts are won generating only 500,000 hours, then selling prices will have to be calculated using £10/hour for direct labour, but this time plus £4/hour for overhead recovery since the reduction in work will not be matched by a proportionate drop in the factory running costs. Thus the selling price goes up and becomes less competitive. A spiral, then, can continue until the firm no longer wins any contracts and has to cease business.

This naturally is one reason that traditional industries in the UK with their massive capital investment and overhead costs have become uncompetitive and slipped into decline. The modern successful firm has minimum capital investment and maximum flexibility in controlling and reducing running costs – for example, by renting rather than buying premises and employing some specialists, accountants, etc., as subcontractors rather than as employees as such.

(8) Contract types

The degree to which cost, time, performance and profit is linked to the nature of the contract in so far as the type of price is concerned will be reflected by the type of contract.

8.1. Firm price

This is the most common type of price – a price which is agreed by whatever means by the two parties to the contract to be the price for all of the work under the contract, regardless of the difficulties and extraneous factors which the manufacturer/supplier may encounter. It is a price which will not vary from the time it is agreed, for any reason. That is, the price for the defined and specified work requirement will not vary. However, if that work requirement is changed by the customer then the supplier is entitled to change the price and indeed the time allowed for performance of the contract.

Throughout this explanation of contract price types it is assumed that the content of the work is not varied by the customer. In the case of the firm price contract therefore the price also does not vary. However, whilst the price is not permitted to vary, the cost to the supplier of undertaking the work may indeed alter to a degree outside of his contemplation.

Within his contemplation may be uncertainty over the accuracy of cost-estimating and he may make an estimating-allowance for this in the price, based on previous experience. He may also consider that this cost may be affected by general inflation and he may make a contingency allowance for this in his price. However, the actual incidence of these known but unpredictable events, and also the incidence of unknown events such as subcontractor industrial-strike action, may be to cause his costs to prove to be highly underestimated or highly overestimated. In the former case the difference between estimated and actual costs will be an erosion of profit or even a loss and in the latter case the difference between estimated and actual costs will become additional profit. This relationship between cost and profit can be seen in Fig 5.2.

Clearly in a firm-price environment the accuracy of cost estimating and the efficiency of contract performance is most vital. The supplier is at maximum risk as he must absorb any and all cost overruns.

Figure 5.2: Fixed price contracts

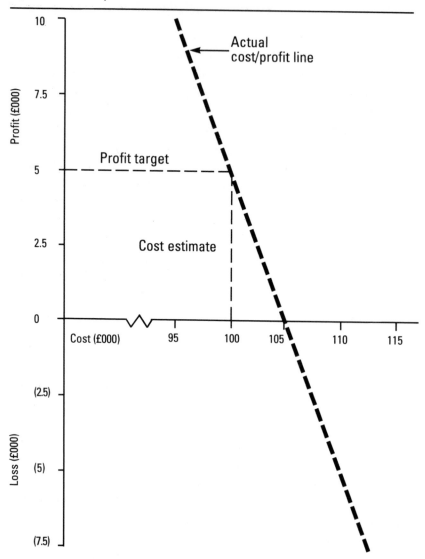

Effects: 1 if costs are £100,000, profit is £5,000;
2 break-even point is £105,000;
3 all costs above £105,000 result in a loss;
4 all savings against the estimated cost of £100,000 will be added to the profit.

'Fixed price is £105,000'

8.2. Fixed price

Fixed-price contracts are very similar to firm-price contracts except that the price is variable based on movements in some factor that influences contract costs. For example, the contract price may be linked to inflation indices recording the change in labour and/or material costs or both. One disadvantage here is that published indices relate to sectors of industry as a whole and the environment in which a particular firm operates may cause him to be more adversely affected by inflation than is suggested by the relevant indices. Another varying cost to which the price may be linked is currency exchange rates. A supplier offering goods or services including a significant foreign content in respect of which he must pay in the foreign currencies will be at risk if his customer wishes to agree the price in sterling only. An element of this risk can be passed on to the customer if he is prepared to include an exchange rate variation clause.

A typical variation of price (VOP) clause covering inflationary effects would be expressed by a mathematical formula of the following type:

$$P_f = 0.1\,P_0 + P_0 \left(0.6\frac{L_f}{L_0} + 0.3\frac{M_f}{M_0} \right)$$

where P_f is the final price, P_0 the basic price for VOP purposes, L_f the final labour index, L_0 the initial labour index, M_f the final material index and M_0 the initial labour index.

In such a formula there is usually an element (here 10%) of the basic price which is not subject to VOP adjustment and a split (here 2:1 making up the balance of 90%) between labour and material indices to recognise both that the movement of inflation in each may be different and that the content of the work may not be evenly divided between the two elements.

The initial indices may be chosen as those applying at the date of price quotation or at some earlier of later time. For example, a firm using September as its economics base-date may offer a quotation subject to VOP where the initial indices are those applying in September although the quotation may be dated December. The final indices may be chosen as those applying at the date of final delivery or at some mid-point if delivery is spread over a significant period of time. Both the initial and final index dates may be different between the labour and material elements to recognise that labour will be deployed over a different time-frame to that in which materials are bought.

The relationship between cost and profit on a fixed-price contract is identical to that in the firm-price contract (Fig 5.2), albeit that the supplier is at somewhat lesser risk, having passed some element(s) of risk onto the customer. In practice, though, the entire risk in relation to that feature may not have been passed on because either there is a non-variable proportion of the price or because the indices on which the VOP formula depends do not fully reflect the effect on the individual firm.

Where the price is adjusted by a VOP clause the operation of the clause can be implemented in a number of ways, depending upon whether payments are being made in advance of delivery or on/following delivery only. If interim payments are being made on a stage-payment arrangement, where the values of stages are percentages of the total price, then the price and thus the stage values may be adjusted as each stage-payment becomes due. Alternatively, where payment is on final completion then there will be one single adjustment of the price and one single VOP payment. As an example, the formula can repeated with some real values:

$$P_f = 0.1\,P_0 + P_0\left(0.6\frac{L_f}{L_0} + 0.3\frac{M_f}{M_0}\right)$$

where P_0 is £100,000, L_f is 180, L_0 is 100, M_f is 130 and M_0 is 100.

The P_f is £157,000 and thus the VOP formula had added £57,000 to the price.

These general principles applying to VOP in relation to inflation also apply to other variable parameters such as exchange-rate variation.

8.3. Cost plus

Quite simply, a so-called cost-plus contract is a contract in which the customer pays the supplier all his actual incurred costs plus something for profit. The something for profit is either a predetermined and agreed sum of money or a predetermined percentage of the actual costs. In neither case is the seller at any risk as he is guaranteed to recover all of his own costs no matter what happens in the execution of the contract, provided of course that he meets the requirements of the contract. Where the profit is a predetermined sum of money the supplier's profit rate as a percentage return on cost reduces as cost increases. Where it is a predetermined percentage then his absolute profit increases as costs increase. These relationships can be seen in Figs 5.3 and 5.4. Neither type of cost-plus contract is inducive to efficiency, cost reduction or timely performance and is usually avoided by customers.

Figure 5.3: Cost-plus percentage fee contract

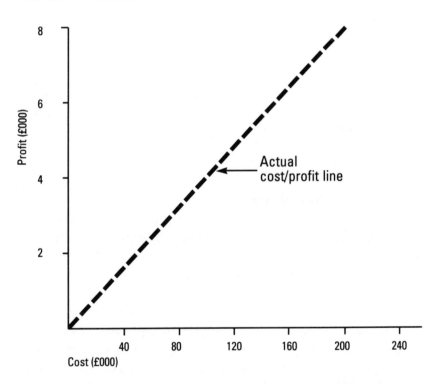

Effect: The seller always earns a profit at a predetermined percentage of cost, irrespective of the level of cost.

Figure 5.4: Cost-plus fixed fee contracts

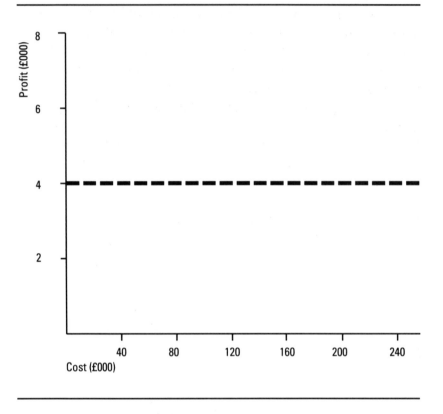

8.4. Incentive contracts

Incentive contracts, as the name suggests, are designed to put pressure on the contractor to strive for efficient performance whether in terms of cost, time-scale, technical performance or combinations of these. In a firm-price or fixed-price arrangement the pressure is on the contractor to perform well, but to do so is in direct terms to his sole advantage. That is, the customer derives no benefit from efficient performance on a fixed- or firm-price contract. He was, after all, entitled to timely delivery of the correct product. In a cost-plus contract there is no pressure on the contractor to perform efficiently, in fact quite the opposite, to the customer's clear disadvantage. The purpose behind an incentive contract is to create this pressure on the contractor but at the same time to allow the customer

Figure 5.5: Fixed-price incentive contract

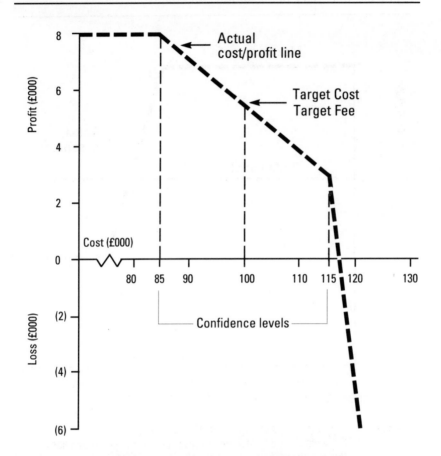

Assumptions: 1 Target cost is £100,000 with fee of £5,000;
2 upper and lower confidence levels are £115,000 and £85,000 respectively;
3 overspends or underspends within the confidence levels will be shared in the proportion 80/20;
4 maximum price is £117,000 with a fee of £2,000.

Effects: 1 For each £5,000 saving against target, the seller's profit increases by £1,000, eg at £90,000 cost he earns £7,000 profit;
2 all costs above £117,000 result in a loss, eg at £120,000 the loss is £3,000.

some concrete benefit from the contractor's efficiency.

Figure 5.6: Cost-plus incentive-fee contract

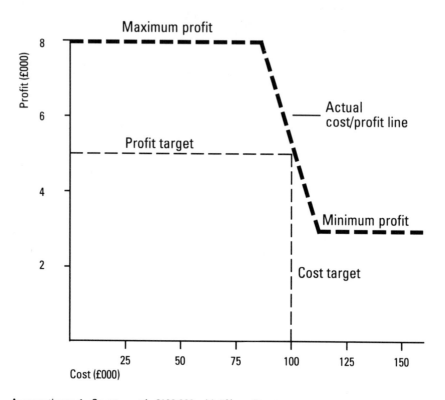

Assumptions: 1 Cost target is £100,000 with 5% profit rate;
                 2 overspends or underspends are shared in the proportion 80/20
                 (buyer/seller);
                 3 maximum profit is £8,000 and minimum profit is £3,000.

Effects: 1 If cost is less than £85,000, all savings accrue to buyer;
                 2 if costs exceed £110,000 the overspend is borne by buyer.

The most common form of incentive contract is that related to cost only. The customer and supplier agree a target cost for the work and agree how each will take a share of any savings or contribute to any overspends against that target. Additionally they will establish a target fee – that is, the sum for profit which the supplier will earn if the actual cost exactly equals the target cost.

In such an arrangement therefore the customer is agreeing to pay the supplier his actual costs plus the target fee, which shall be increased or decreased as appropriate by the agreed share of underspends or overspends against the target cost, their respective shares in either case being known as the share ratio and agreed in negotiation. Conventionally the customer will take the lion's share of underspends and overspends. A typical example is illustrated in Fig 5.5. This demonstrates the 'cost incentive range', being the span of cost between the most optimistic cost and the most pessimistic cost, two figures also agreed in negotiation between the two parties. In this example it can be seen that the supplier can earn unlimited profit on the underrun side. Countering this on the overrun side is the maximum-price feature – that is, the extrapolation of the share line beyond the most pessimistic cost shows the erosion of the supplier's fee on a 0/100 ratio until excess costs have eliminated all profit and the supplier goes into a loss.

It is open to the parties to negotiate different incentive schemes. An alternative is shown in Fig 5.6. In this situation there is no maximum price, which has the effect of saying that no matter what the costs are, the supplier will receive a minimum fee. The quid pro quo for this is that there is a limitation on the maximum fee that the supplier can earn.

In addition to benefits to both sides relating to the supplier's performance in meeting or bettering the target cost, the parties may agree that the supplier can earn an additional fee (i.e. over and above the target fee and shared cost savings) for early delivery or improved technical performance. As far as delivery is concerned, the scheme may be quite simple, e.g. for each week that the supplier delivers ahead of schedule he shall earn the sum of £X. This is relatively straightforward to plan and monitor. Technical-performance incentives in quality standards, reliability or specific features such as power-to-weight ratio are more difficult to achieve and measure.

In a mixed-incentive scheme there is a complicated trade-off analysis to calculate. Extra cost means less savings to be shared but may allow higher technical performance, which will earn its own enhanced profit. Early delivery may bring extra profit but at the sacrifice of not introducing performance improvements, thus missing other profit opportunities.

Price make-up

Regardless of whether the price is to be a budgetary estimate or quotation (firm, fixed, cost-plus or incentive; established by competition, advertisement or negotiation) it will have the same basic elements in its make-up (see, for example, Table 5.1).

Table 5.1.

Direct labour: 100,000 hours at £15/h =	1,500,000
Overhead recovery, 20% of above	300,000
Subcontract work	200,000
Materials	500,000
Basic works cost =	2,500,000
Risk allowance	
Design: 10% of above	250,000
Manufacturing: 5% of above	125,000
Subtotal	2,875,000
Special to contract allowances	
Liquidated damages protection	25,000
Warranty	250,000
Cost of financing work in progress	100,000
Subtotal	3,250,000
Profit @ 10%	325,000
Selling price	3,575,000

So in this classic price make-up all the elements referred to earlier in this chapter are included to the extent appropriate. The direct labour, subcontract and material estimated costs, all wholly attributable to the contract to which this price relates, are shown. The overhead recovery expressed as a percentage of the direct labour is there to recover this contract's proportionate share of the factory set-up and running costs. The risk allowances are either standard percentage mark-ups or calculated by reference to an estimate of what is reasonably likely to go wrong and the cost of putting it right, the base to which the risk allowances will vary according to circumstances.

In the example shown the design-risk allowance has been applied to the subcontract element because some of the design work is to be done by subcontractors. On other occasions the design-risk allowance may be applied to the direct-design-labour only, similarly for the manufacturing risk. The other categories of risk mentioned earlier in this chapter could each be estimated and specific allowances made. The practice is to make specific allowance for major risk areas only, although quite commonly an overall contingency factor may be applied to cover risk in general.

The special-to-contract allowances are in respect of those features unique to the particular contract which imply cost or risk. Special warranty requirements, perhaps a warranty period of twice the usual time, implies both cost and risk. If the contract includes a liquidated-damages clause then the supplier is wise to include in his price an allowance for having to pay the customer for his lateness. These are only examples of course. It is vital that the contract is thoroughly scrutinised for any features that imply cost or risk.

The final element in the price build-up is profit. Many companies have various standard profit mark-ups, depending on the product, market and customer.

In a fiercely competitive environment it is necessary to match the ideal price build-up, with all its allowances and contingencies, to whatever intelligence there is as to the likely winning price. It is no good having a safe price in terms of full coverage of all costs, risk and profits if it is not a winning price. It should be remembered that the difference between the basic works cost and the selling price is the buffer or margin between profit and loss. It is assumed that the basic works cost is the minimum figure for which the work can actually be done. The selling price is the maximum which the customer will pay. The sum included for profit is notional only. Efficient performance with elimination of the perceived risks will convert the whole margin to profit. Inefficient performance and ineffective risk management can erode the margin to zero and plunge the contract into loss. In the example the profit opportunity is as much as £1,075,000, being a return on cost of 43% compared with the notional 10%.

Different companies have different cost-accounting systems and thus each will have differing treatments of categories of cost and handling of risk, but the foregoing example serves to illustrate the principles involved.

The only other factor not mentioned in the foregoing but which sometimes is applied is that of a negotiating margin. This is an additional sum included in the price which can, in whole or in part, be eliminated in negotiation. This building in of something to give away, so that the real allowances and margins are protected and preserved, is a feature to which the customer will be alert.

6

CHAPTER

Getting paid

1 Introduction

In Chapter 3, cash features amongst the list of key commercial factors. Cash means getting paid. Therefore, whether in pre-contract negotiations or during contract performance, there is a need not to lose sight of the arrangements surrounding the payment provisions.

The four elements to payment are:
1) The customer's obligation to pay.
2) Establishing that payment is due.
3) The mechanism for invoicing.
4) Timing.

In this section some of the practical issues concerned with these four aspects are examined.

2 Obligation to pay

As a matter of law the customer is obliged to pay, provided that the seller has performed his obligations under the contract. Given that the customer is obliged to pay, the question is: what does he have to pay? Payment will be the price of the goods or an element of the price if the payment is interim, i.e. made before contract performance is complete. The contract will specify the currency or currencies of payment. Even in the UK payment may be in mixed currencies if a significant proportion of the work is subcontracted overseas. If the seller has to pay the overseas supplier in local currency then the seller will prefer the buyer to pay him in that currency for that part of the work. This moves the exchange-rate-variation risk from the seller to the buyer. The buyer may be quite content to adopt this arrangement if he is able to buy the particular currency more cheaply than the seller, or if to do otherwise is to accept a higher price from the seller as the seller builds into his price a contingency to cover the exchange-variation risk. The contract must also specify the applicable taxes and other charges that the buyer is obliged to pay. For example, payments may include VAT, import duty, carriage and insurance charges.

If the contract allows interim payments, these will generally be on the basis of a regular reimbursement of costs actually incurred or payments of specified sums or specified percentages of the contract price. In the latter case the sums are payable on completion of predetermined stages of the work. The two approaches can be combined. For example, a contract of long duration in which it is not possible to define concrete stages for the early part of the work may permit regular cost reimbursement (progress payments) until the work becomes susceptible to stage definition.

The stages or milestones for a contract for development and supply of the Mark 200 widget might, for example, be as shown in Table 6.1.

Table 6.1.

Stage	Milestone definition	% of price	Cumulative
1	Performance specification	5%	5%
2	Final design review	5%	10%
3	Production drawings	5%	15%
4	Materials	10%	25%
5	10% subassemblies	5%	30%
6	90% final test	30%	60%
7	1st delivery	20%	80%
8	Final delivery	20%	100%

In practice, the definition of these milestones must be spelt out in full so that there is certainty as to what achievement of the milestone actually means. The seller will try to make the milestone definitions as soft as possible, the buyer as hard as possible. For example, the seller will propose that milestone 1 is defined as the issue of the performance specification, the buyer will argue for acceptance of the performance specification. Milestone 2 will be defined by the seller as the occurrence of the final design review. The buyer will look for something stronger, such as acceptance by the buyer of the minutes of the final design review. The buyer will negotiate milestone 8 as acceptance of the final delivery rather than just the physical event. In the extreme the buyer might wish to retain some payment until the expiry of any warranty period or the satisfactory repair of any warranty failures. As well as the respective aims of the buyer and seller it is as important to ensure that the definitions are sufficiently detailed. For example, milestone 3 might be more fully defined as production drawings issued to the shop floor or approved by the quality manager.

The milestone scheme may also include a column to indicate the planned or forecast date for completion of each stage. It is important in these circumstances to refer to planned or forecast dates so as to avoid the possibility of a legitimate payment being withheld because, although the stage was achieved, the indicated date was not the date on which the event occurred.

Each stage may have more than one milestone associated with it and a separate payment. For example, a scheme might include the stages shown in Table 6.2.

Table 6.2.

Stage		Milestone	% of price
23	a)	Drawings frozen	3%
	b)	Tool-setting complete	2%
	c)	Components loaded	4%
24	a)	Frames drilled	1%
	b)	Corrosion treatment complete	2%
	c)	Sub-assembly inspection complete	3%

The buyer ideally would like stage 23 to describe drawings frozen and tool-setting complete and components loaded before he is committed to pay anything for stage 23. The seller will wish these to be and/or provisions so that, even if one or more milestones in the stage are missed, there may nevertheless be some payment due. This is particularly so where there are constraints on the availability of payments: for example, if payment under the contract will not be made more frequently than quarterly, or where it is not possible to plan in advance the exact timing or sequence of events. The more milestones and stages there are the better it is for the seller, who can forecast receiving regular payments rather than risk all on one or two major milestones.

③ Payment due

The foregoing section has discussed the details that in practice describe the obligation to pay. From this the question of the point at which payment becomes due follows naturally. The four elements in this are:

1) What evidence is necessary to support a claim that payment is due?

2) Are there any time-constraints on payment?

3) Are there other conditions of the contract that must first be satisfied in order for payment to be made?

4) Can the buyer withhold payments due?

The evidence necessary depends on the type of payment arrangements. If the contract provides for payment on delivery then the evidence is the appropriate delivery documentation signed by persons authorised in the contract. If progress payments are to be made the evidence will be a certificate signed by the seller's accountant, finance manager or other authorised body, to the effect that the costs claimed have actually been reasonably and properly incurred.

Where the contract is subject to stage payments then documentation must be produced as evidence that stages claimed have been completed. These stage-completion certificates may be acceptable evidence if signed by an authorised officer of the seller, or the buyer may wish himself to physically verify that the stage is complete. In the latter case the buyer may seek to avoid a stage scheme with too many milestones because of any inconvenience in verification.

There may be some time-constraints on payment becoming due. If the buyer's own availability of funds is limited he will wish to include in the contract provisions to limit his actual expenditure to pre-set sums being available not earlier than pre-set dates, regardless of achievements under the stage-payment scheme. He may include, for example, in the contract words to the effect that 'in any event payments shall not be made in advance of the following schedule:

£50,000 in fy 90/91

£150,000 in fy 91/92

£75,000 in fy 92/93.'

The second possible constraint on timing of payment is the likelihood of the buyer reserving himself time in which to make payment even when a delivery or milestone stage is complete and an invoice submitted. This is quite legitimate in so far as the buyer must quite reasonably be permitted the time and opportunity to verify that all is in order prior to making payment. In practice the period can be quite long, ninety days is not unusual, and the length of the period is almost entirely a function of the respective bargaining positions and strengths of the two parties.

In practice the contract may also impose other preconditions for payment to the seller. For example, the buyer, if he is not the end-user, may say that he cannot pay until he has been paid by his customer. To the seller this is most unreasonable as he sees only the contract between himself and his buyer and the buyer's other affairs are no concern to him. The buyer on the other hand has only placed the contract to help him satisfy his customer's requirements. The usual compromise is to line up the seller's claims for payment with the buyer's claims from his customer and to agree a long-stop whereby, if the buyer does not receive payment from the customer through no fault of the seller, then the buyer will in any event pay the seller after the expiry of, say, sixty days.

The buyer may in any event have discretionary or prescribed powers to withhold payment, or indeed to recover payments already made. He may wish to possess the power to make interim payments at his sole discretion or to have the right to withhold or recover payments for any default or deficiency on the part of the seller or if, in his opinion, the seller is not making positive progress with the work. Unless he is in a superior bargaining position there is little the seller can do to resist the buyer's objectives in this regard, although he may be able to secure agreement that there must be a period of notice before the buyer's powers to withhold or recover payments can be implemented and that he be given a reasonable opportunity to cure whatever is the problem.

4 Mechanism for invoicing

Having established that the buyer is obliged to pay and that payment is due, it is equally important to ensure that there is a clear understanding of the mechanism for payment. Armed with whatever evidence of payment due is necessary, the seller must be sure that his paperwork for invoicing is correct and in line with the contract. Errors in the customer's name and address or contract number or in the details of the stage or milestone or delivery for which payment is claimed can lead to the invoice being rejected. The bill must therefore be re-submitted, which risks further delay in payment being made. the invoice must be correct in all its details; currency, VAT amount, VAT registration number are other potential areas for error.

5 Timing

From the foregoing and Figure 6.1 it can be seen that payments are subject to:

a) Making progress with the work according to any time plan.

b) Producing the right evidence of payment due.

c) Any funding availability restraints of the customers.

d) The customer's rights to delay payments for a period of time.

e) The customer's rights to withhold or recover payments.

f) Invoice rejection on the grounds of error or incompleteness.

Figure 6.1: Payment

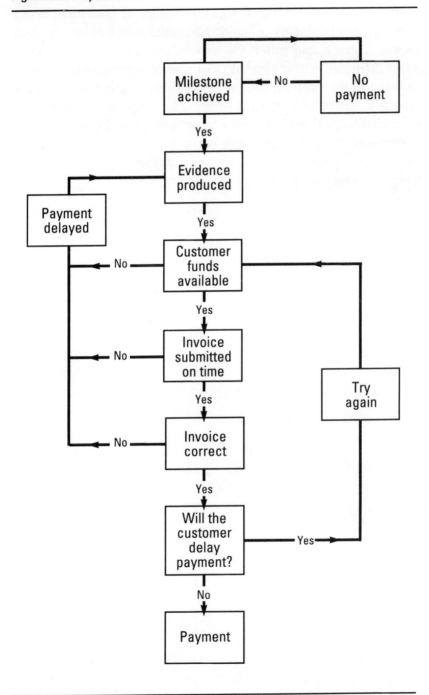

But against this background the seller must make his predictions of cash intake so that he can make his arrangements accordingly in terms of his payments to employees and subcontractors and other disbursements.

If the seller has agreed to quarterly stage-payments with his customer and with the customer having the right to delay payment, there can be great problems if employees are weekly or monthly paid and his own suppliers must be paid within twenty-one days. Clearly each party vitally needs to negotiate the optimum payment terms with customers and suppliers to ensure a stable and safe position.

7

⬤ ⬤ ⬤ ⬤ ⬤ ⬤ ⬤ ⬤ Ⓒ Ⓗ Ⓐ Ⓟ Ⓣ Ⓔ Ⓡ ⬤ ⬤

Delivery

① Introduction

The simple word 'delivery' cannot be considered in a commercial sense without addressing the question of who has the goods, who owns them, who carries the risk if they are lost or damaged and who is liable if they do not work.

② Possession, ownership and risk

2.1. Definitions

Possession, ownership and risk of loss or damage to the goods are bound up together. Let us consider the definitions:

Possession This is concerned with the physical location of the goods. Does the buyer or seller or a third party actually hold the goods.

Ownership In legal terms, ownership is known as property (as in 'the property in the goods vests in X') or title (as in 'X has title to the goods'). Property and title can have different meanings, but for these purposes it is adequate to consider them as having the same meaning.

Risk This is concerned with who is liable for loss or damage to the goods. It is worth mentioning the distinction between liability and responsibility. If the goods are stolen the seller might be liable to the customer for their replacement albeit that the thief was responsible for the loss.

Before developing the relationship between possession, ownership and risk it is necessary to explain two important background principles. Firstly, it must be presumed that the seller has the right to sell the goods. It is not possible to sell things that do not belong to you unless you are acting as agent for the owner. Thus a 'buyer' of stolen property gains no ownership of the goods as the 'contract' of sale is not enforceable. Secondly, it is also presumed that the seller is entirely free to sell the goods. The innocent buyer will not be aware of the existence or details of any other contract that the seller holds, any of which may entitle another customer to have a legal claim on the seller's property, which could include the goods that the buyer believes are for him. So for the purposes of this discussion it is assumed that the seller owns the goods he intends to sell and that there is no legal encumberance to prevent him from doing so.

2.2. Delivery

It might be thought that the definitions given above should have included a definition of delivery. However, it is not possible to come up with a general definition. Commonly delivery is concerned with:

a) the timing of the goods being available to pass into the buyer's possession,

b) the place where the buyer will take possession,

c) any necessary paperwork.

It would be pleasantly simple to define delivery as the event by which the buyer concurrently takes possession, acquires ownership and accepts liability for loss or damage. However, very frequently these three events do not happen concurrently.

The Sale of Goods Act 1979 recognises this situation where it says that: 'Unless otherwise agreed, the goods remain at the seller's risk until the property in them is transferred to the Buyer, but when the property is then transferred to the Buyer the goods are at the Buyer's risk whether delivery has been made or not.'

2.3. Acceptance and rejection

There is one other factor to consider in this subject and that is acceptance/rejection. There is a point in time when the buyer, whether by his actions or by compliance with a requirement of the contract, must have accepted that the goods are what he wanted and that his right to reject the goods for not being what he wanted has disappeared. In practice 'what he wanted' means that for which he contracted, and the two may not be synonymous. In practice, the period available to the buyer to reject the goods can be a specific number of days or weeks, or if none is specified than a reasonable time, having regard to all the circumstances. If the contract includes detailed arrangements and procedures for the goods to be accepted than there could be no right of rejection once the acceptance procedures are satisfactorily completed.

2.4. Buyer and seller

As regards ownership and risk, the positions of the buyer and seller are diametrically opposed. The buyer will want to acquire ownership as early as possible and risk as late as possible. Generally the seller will wish to retain ownership until the last possible moment but to pass the risk to the buyer as quickly as possible. There are two general observations that can be made:

a) Ownership will pass to the buyer when payment is made. Thus if payment is made on delivery, ownership passes on delivery. It is, however, open to the parties to agree that ownership can pass at different times. For example, if the buyer makes advance or stage payments prior to deliveries being made, then ownership can pass prior to delivery. If payment is due on delivery but in practice is delayed because some other procedural aspect of the contract has not been satisfied – perhaps where payments are made once a month only on a predetermined day – then ownership may nevertheless pass on delivery. Thus the aim is to have ownership pass on delivery, but in practice regard will be given to the timing of payment. Again, the principle is that delivery does not have to be concurrent with the buyer taking physical possession. Deliveries can be made on a 'self-to-self' basis, with ownership passing to the buyer but the goods remaining with the seller to hold for collection or perhaps to utilise on some other contract with the buyer – embodiment into other goods for example.

b) Risk will pass when the parties agree. Trading practices exist on which the parties may rely. For example, contracts may be 'ex-works', 'Free on Board (FOB)', 'Cost Insurance Freight

(CIF)', which have generally accepted meanings, of which more later. Clearly, again the question of possession is important, but not invariably so. The party with possession may be best placed to take care of the goods and therefore to assume liability for loss or damage. However, if the buyer rejects the goods after taking possession he may not wish to carry this liability as the goods are no longer of any interest to him. Similarly, if the goods are found to be defective and the seller is liable under warranty, the buyer may not acknowledge a contractual liability to take the risk of further loss or damage. If the contract is cancelled after part-delivery does the buyer have the risk in the goods already delivered?

2.5. Delivery terms

The consideration of possession, ownership, risk, rejection, acceptance, warranty can be seen in Table 7.1.

Table 7.1.

	Manu-facture	Delivery	Period for rejection	Period for warranty
Posses-sion	Seller	Buyer	Buyer until rejected goods returned to seller	Buyer until defective goods returned to seller
Owner-ship	Seller (but Buyer if interim payments are made)	Buyer	Buyer (or seller if the contract permits)	Buyer if goods are to be repaired. If goods are to be replaced, ownership may revert to seller
Risk	Seller (even if prior payments made)	Depends on contract type (see fig. 7.2)	Buyer (or seller if the contract permits)	Buyer if goods are to be repaired. If goods are to be replaced risk may revert to seller

As far as delivery is concerned, the point at which risk passes depends on the type of contract. The types range from 'ex-works', where the responsibility of the seller is limited to making the goods available at his premises for collection by the buyer, to 'delivered duty paid', where the seller is responsible for delivering the goods overseas to the ultimate destination with all costs and duties paid by the seller (but included in the selling price of course). The former type is minimum risk to the seller and maximum risk to the buyer, these positions being reversed in the latter type. The full range of contract types is illustrated in Table 7.2. Where the buyer or seller employs a third party as carrier, the risk as seen by the contract between buyer and seller does not change.

Table 7.2 Passing of risk

	Risk assumed by buyer at:	Seller risk	Buyer risk
Ex works	Seller's premises	Min.	Max.
Free carrier (named port)	Carrier taking custody		
FOR/FOT (Free on rail/ free on truck)	Carrier taking custody		
FOB/airport (Free on board/airport)	Carrier taking custody		
FAS (free along side)	Goods placed on quay		
FOB (free on board)	Goods pass ships rail		
C & F (cost and freight)	Goods pass ship's rail		
CIF (cost, insurance and freight)	Goods pass ship's rail		
Freight carriage insurance	Carrier taking custody		
Ex ship	On board at destination		
Ex quay	On quay at destination		
Delivered at frontier	Frontier at country of destination		
Delivered duty paid	At ultimate destination	Max.	Min.

The employer of the carrier will no doubt contract with the carrier on the basis that the risk is transferred to the carrier whilst the goods are in transit. However, in, say, a CIF contract where the seller engages a carrier, the buyer will hold the seller liable for loss or damage, albeit that the seller may take action against the carrier.

2.6. Documentation

One practical issue of the utmost importance is the paperwork associated with effecting delivery. Many deliveries are affected by a system of 'advice and inspection' notes. These may comprise a delivery note and a certificate of conformance. The delivery note must be in strict accordance with the contract. If the delivery is 10 of Widget Mk 3 then the delivery note must not, for example, refer to Widget Mk 2 unless the customer has confirmed that Mk 2s are acceptable, in which event the contract should be amended accordingly. Similarly with the certificate of conformance. Whether this is a document to be signed solely by the seller or countersigned by the buyer, the point remains that it is a certificate to the effect that the delivery being offered is strictly in accordance with the contract.

③ Defects after delivery

3.1. Introduction

At the point of delivery, legal entitlement to the goods passes from the seller to the buyer. The seller ideally would like this to be the absolute end of his responsibility. The contract has been performed, the goods are off his site and the customer has taken delivery. On delivery the supplier will usually 'take' his profit, this is an accounting convention that says the risk to the seller has disappeared and therefore it is safe to take the profit – the difference between cost and price – into his books. Further work will simply erode the profit margin. Nevertheless the buyer is entitled to expect that the goods will work satisfactorily for a reasonable period of time and that any defects found after delivery will be put right at the supplier's cost. The buyer's rights in this regard can be established in several ways:

a) By including in the contract a specific condition relating to liability for defects found after delivery. This is known as an express warranty.

b) By relying on statute. The principle statute being the Supply of Goods Act 1979. This is known as implied warranty.

c) By relying on common law regarding negligence.

3.2. Express warranty

The buyer in principle will want the supplier to carry unlimited, unbounded and unending liability to correct defects. The seller will naturally wish to limit his warranty obligation and to minimise the cost of corrective action. This he will pursue on five fronts:

a) A warranty period.

b) Limitation on what is covered by the warranty.

c) Limitation on the circumstances in which a defect counts as a warranty claim.

d) Actions the buyer must take on discovering the defect.

e) Limitation on the actions that the seller must take to correct the fault.

a) The warranty period. There are two key issues relating to the warranty period: one is the length of the period and the other is the start-point. The length of the warranty period is highly variable and appears mostly to depend on custom and practice in the particular trade or sector of industry. The seller may offer his 'standard' warranty period or the buyer may wish to specify a period depending on his unique needs. It is nowadays common to have three-year warranties on household electrical goods and up to five years on motor-vehicle bodywork where once 12 months would have been normal. The warranty period can be a powerful selling or marketing point as it not only gives protection to the customer but also gives him reassurance (and therefore encouragement to buy) that the product is safe and reliable. The start-date for the warranty can begin at a number of points, e.g.

a) The time of physical delivery to the customer.

b) The time of final acceptance by the customer.

c) The time at which the customer takes the goods into use.

The time of physical delivery invites consideration of when and how that takes place. If the contract is for 50,000 widgets then most sensibly the warranty applies to each widget and the period for each starts on physical delivery of each. Thus a two-year warranty applying to 5,000 widgets to be delivered over three years will leave the seller with the cost of warranty repairs potentially for five years. On the other hand, if the contract is for 3 mega widgets perhaps it is more relevant to commence the warranty on delivery of the final unit.

Physical delivery is one thing, but there may be a delay before the customer takes the goods into use. Clearly the supplier would like the clock to start ticking on physical delivery, the buyer when he starts to use the equipment. A compromise here would be to agree that the warranty is two years, commencing from delivery or 12 months from the date of taking into use, whichever period expires the sooner. The final situation is one in which there are extensive contract acceptance or handover procedures. This may be where the nature of the goods – perhaps a power-station or battlefield communications network – cannot be fully tested and stressed by the supplier before delivery. In that event the warranty should commence at the end of the acceptance or handover phase. This leaves the supplier with the problem of beginning his warranty period possibly well after the goods have actually been taken into use.

b) Limitation of coverage. The limitation on what the warranty covers can be defined in several ways:

a) By warranting the design.

b) By warranting the product but only in so far as defective materials and/or workmanship are concerned.

The customer, of course, will not wish to see the warranty limited in these ways. He is only concerned as to whether the goods work or not. If not, the customer is not on the face of it interested in whether the design or the material or the workmanship was at fault. In practice this could be of paramount interest to the customer. Defective workmanship may mean only one unit is faulty, defective materials may mean that an entire batch is faulty and defective design may mean that all past production is faulty. Car manufacturers buying brake components from component suppliers will be keenly concerned, as their warranty obligations to the end customer, the car buyer, may be to rectify the car, incurring labour and material costs where perhaps only the liability for material costs can be passed back to the component supplier.

c) Limitation of circumstances. Limitation of the circumstances in which defects count as a warranty can be achieved by excluding defects such as the following:

a) Those caused by incorrect storage, operation or maintenance.

b) Those caused by fair wear and tear.

c) Those not notified promptly to the supplier.

d) The buyer's unauthorised modifications.

Earlier in the text the point of commencement of warranty was discussed. In some circumstances the customer may take delivery well before putting the article into use. If the goods are not stored properly during this period and defects are thereby induced, then it would be unfair to expect the supplier to accept warranty liability for those defects, always provided that the supplier made available adequate storage instruction, assuming that he knew that the customer would delay taking the goods into use. Incorrect operation could similarly excuse the seller from liability, again provided that the goods had been supplied with adequate operating instructions. There is a grey area here in so far as reasonable behaviour is concerned. If the operating instructions for a domestic stereo system do not specifically say that the equipment should not be dropped on to concrete from a height of 6ft then a defect arising from being dropped could not reasonably be a warranty claim. On the other hand, if switching on the power with the volume control set at maximum causes damage to the speakers, is it reasonable for the owner to call this a warranty defect if the instructions did not warn against this hazard? Possibly. The message really is that operating instructions must be as comprehensive as possible, although this may have some other disadvantage if, for example, customers do not find lengthy operating instructions attractive. Similar concepts apply as regards maintenance. Some products (motor vehicles, for example) carry warranties that are invalidated if maintenance is not carried out by an authorised agent in accordance with the service schedules. Some domestic electrical-goods warranties are invalidated if the customer so much as takes the back off. Where the customer is permitted to carry out his own maintenance (this can be a matter of policy, as with the armed forces) during the warranty period the maintenance instructions will need to be comprehensive and may need to go so far as specify the qualifications or skills of the maintenance engineer and the equipment and facilities that need to be used for maintenance.

It may be necessary also to exclude from warranty fair wear and tear. This could include such things as components or modules where the technical performance is such that failures are predicted after certain running times ('Mean Time between Failure'). Provided these are accepted by the customer as a limitation of the equipment's performance, then they could not be deemed to be within the scope of warranty.

As a matter of fairness, the buyer ought to promptly bring to the attention of the supplier defects as they arise. Clearly it is unreasonable for the buyer to notify defects 12 months after they have occurred. What is a reasonable period of notice is somewhat difficult to decide. A major factor will be the nature of the goods. At the end of the day it will be a matter for negotiation. The supplier may want immediate notice, but this may not be practicable or acceptable to the customer.

Finally, the supplier will not usually accept any warranty liability if the buyer makes modifications to the goods. That is, the warranty condition becomes invalid whether or not the defect was caused by the unauthorised modification.

d) Obligations on the buyer. To minimise the seller's exposure he may wish to impose obligations on the buyer to mitigate the effect of any defects arising. This could be limited to the effect that the buyer must mitigate effects or to comply with specific instructions set out in operating instructions or given by the supplier at the time he notified of the defect. The mitigating actions could range from undertaking some minor engineering process to ceasing operation of the equipment. In some cases, such as a television catching fire, the customer in any event is not going to carry on using the equipment. However, a defect (not causing total failure) to a Tornado radar is most unlikely to cause the RAF to cease flying the aircraft or using its radar. The buyer may seek to negotiate with the supplier the acceptance of liability not only to remedy warranty defects but to reimburse the buyer the extra costs that may arise from being deprived of the use of the goods.

So there may be limitations on the period of and coverage of the warranty, the circumstances in which the warranty can apply, and the obligations that can be placed on the buyer to mitigate the effect of defects.

e) Obligations on the seller. The question then arises as to any limitation on the actions that the supplier is required to take to correct the defect. The buyer will wish to have the option to decide whether:

a) The defective item should be repaired, modified or replaced.

b) The remedial action will be undertaken on site or by return to the supplier.

c) Whether *in situ* remedial action can be undertaken by the buyer or seller.

These are undoubtedly the decisions to be made, and not unnaturally the supplier will wish to reserve the decision making responsibility to himself. If he can succeed in this then he may wish to go one step further and state that the nature of the remedial action will be his to determine. For example, a design defect on a printed circuit board may be capable of correction by a 'cut and strap' modification. The customer may prefer a new board with redesigned art-work.

f) Warranty of repaired/replaced items. The general rule is that repaired/replaced items benefit only from the un-expired period of the original warranty. It is open to the customer to negotiate something different, but the supplier will do his best to avoid full warranty applying from the date of repair/ replacement as this could be a completely unlimited obligation.

Recovery of the costs of warranty repair/replacements will be within the selling price of the goods. For a standard product with a standard warranty the supplier will monitor the type, frequency and cost of repair/replacement and build a suitable allowance into his selling price. Where the product is non-standard, i.e. developed or tailored to the customer's particu-lar requirements, the selling price will still include an allowance for warranty. In negotiating the warranty condition the customer will have regard to the direct link between the terms of the warranty and the cost that will be built into the price. One way of describing this process by which the customer pays for his warranty benefits would be to say that it is a form of insurance policy. For a fixed sum of money (within the product price) the

Figure 7.1: Express warranty

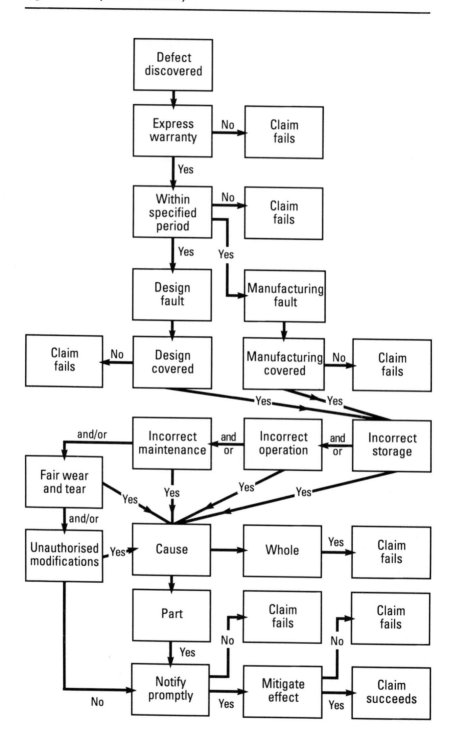

customer passes the risk of correcting failures, for a period of time, back to the supplier.

A possible flow diagram of an express warranty claim is seen in Figure 7.1.

3.3. Implied warranties

Whether or not an express warranty is provided by the supplier and therefore allowed for in the price or paid for specifically as an 'extra' by a customer with particular requirements, the law nevertheless provides some protection as regards the quality of the goods. The 1979 Sale of Goods Act provides the buyer the right to goods that are of merchantable quality and are fit for their purpose. These rights only apply to sale of goods within the meaning of the Act. Thus the Act may not apply to, for example:

a) Contracts of exchange.
b) Gifts.
c) Hire-purchase.
d) Contract for skill and labour.
e) Contracts for labour and materials.
f) Agencies.

Also, the obligation relating to mechantable quality does not apply:

a) As regards defects specifically drawn to the buyer's attention before the contract is made.

b) If the buyer examines the goods before the contract is made, as regards defects which that examination ought to have revealed.

Certain aspects of the Sale of Goods Act were specifically drafted to protect the consumer. The consumer may – wittingly or unwittingly – rely on the Act as regards merchantability and fitness for purpose. However, in non-consumer contracts it is open to the parties to agree that these provisions of the Act do not apply to the particular transaction. Thus in a non-consumer contract where the buyer wishes for an express warranty the supplier may offer this on the basis that it is in lieu of any and all implied warranties. On the other hand, where consumer goods come with a written express warranty, these days there is usually a notice saying that the rights provided under the express warranty are in addition, and do not affect, the rights available under the Sale of Goods Act. Implied warranties, as express warranties, convey a duty to the buyer to mitigate the effect of defects.

3.4. Negligence

The purpose behind express and implied warranties is to afford the buyer some certainty that the quality of the goods will be such that he is not deprived of the use of the goods through defects discovered after goods have been accepted, and to provide some remedy if this does occur.

However, if a defect occurs and the result is more than simple deprivation of use, causing loss or damage to property or land or death or personal injury, then the buyer may also be able to pursue the manufacturer, or in certain circumstances the seller or other intermediary, for negligence.

In most cases the buyer would have sufficient remedies available to him under breach of contract or breach of warranty.

4 Cancellation

A contract can come to a premature end for basically two reasons only. Firstly, the failure of one party to perform his fundamental obligations under the contract permits the other party to repudiate the contract and sue the defaulting party for damages in a court of law. Secondly, the contract, if it contains appropriate provisions, can permit the cancellation of the contract for the convenience of one party. The former remedy is a fundamental of contract law, the latter can only apply if the contract says so.

Failure to perform a contractual obligation is known as breach of contract or default. A minor breach does not permit the contract to be cancelled by the injured party. Generally, the fundamental obligations are for the seller to perform the contract and the buyer to pay. Thus the buyer's fundamental breach is non-payment, and whilst the consequences for the seller could be serious, the breach is nervetheless simple in nature. Provided the time and mechanism for payment is clear then non-payment is readily demonstrable. For this reason the seller may rely on

the court to order payment, provided it is clear that payment
is due. The seller is unlikely therefore to want to cancel the contract since there would then be no contract and therefore no
question of a payment being due. He would be left with the
goods, which may or may not have some other sale value, and
he would have to sue for consequential damages. The seller does
not want to cancel the contract; what he wants is for the court
to enforce the contract and make the buyer pay. On the other
hand, if the seller defaults by delaying performance or failing
to perform then the buyer may reach the stage where he
considers it quicker and more reliable to go elsewhere. In these
circumstances he most certainly will wish to cancel the contract
and remove himself from any liability to pay the defaulting seller.
The buyer may then go on to sue the seller for costs arising
from the default, particularly any extra costs resulting from
going elsewhere.

The difficulty for the buyer in actually taking the seller
to court to recover damages is that it is a costly and time-
consuming business, particularly if it is necessary to prove that
the seller's lateness was such that the buyer's needs could not
wait or that the evidence was such that the buyer reasonably
could not expect that the seller was ever going to perform. This
need for proof might arise, for example, if the buyer had not
specified that time was of the essence of the contract.

If the nature of the work, perhaps a lengthy development/
production package contract, is such that, for the first year or
so of the contract, progress is more a question of opinion than
hard fact, again the buyer might have the onus to prove that
the seller was in default. Given these difficulties, the buyer may
seek to negotiate a specific contract condition giving him the
right to cancel the contract if, in his opinion, the seller is in default, to recover any payments so far made, to go to another
supplier for the goods, and charge the seller for any extra costs
arising. On the face of it, this approach would save the buyer
the cost and trouble of going to court. In practice, the seller could
simply ignore the condition and wait for court action if the
situation ever arose. The seller would perhaps only risk this if
the buyer was not a key customer. If he were a key customer
the seller might accept the condition but negotiate detailed

procedures to apply where the buyer suspects default. For example, the buyer may be required to give notice to the seller of the alleged default and permit a period in which the seller has the opportunity to cure the failure.

The concept of cancellation for convenience is somewhat different. There is no common or statute-law right to cancel for convenience; it can only be provided by a specific condition in the contract. As with cancellation for default, the main interest lies in the buyer's potential desire to cancel. The seller's needs are to sell goods to make profit, and it is generally presumed that the seller would not wish to cancel for convenience. This is not impossible of course, as for example the seller might find an alternative customer for the same goods but at a higher price. In these circumstances the seller might in practice try to negotiate a later delivery with the first buyer. In any event, it is most unlikely that the buyer would give up his right to the goods for the convenience of the seller. In contract the buyer is buying to fulfil a specific need, as opposed to the general need of the seller. Therefore it is far more likely that the buyer will require the right to cancel for convenience as his specific need may disappear.

Where the two parties agree to a cancellation-for-convenience clause, a period of notice will usually be included. The seller may attempt to negotiate that the balance (i.e. as at the time of notification of cancellation) only may be cancelled and that payments must be made for work done. The seller will also be looking for recompense for loss of anticipated profits on the cancelled contract or part contract.

As with many contract conditions, the form of cancellation for default or cancellation for convenience condition will largely depend on the negotiating strengths and positions of the two parties to the contract.

⑤ Customer-furnished equipment data

Frequently the successful performance of the contract will require the customer to provide the seller with equipment and/ or information. Equipment may be required under three main arrangements:

a) Embodiment loan. As the name suggests, this would be where the customer provides equipment for the seller to embody in the goods during manufacture. This could arise where the customer, for a variety of reasons, is able to buy component parts or sub-assemblies more cheaply than the seller, or where the seller is unable to buy the component parts.

b) Contract loan. The concept here is that items are loaned for whatever purpose is the actual subject of the contract. For example, in a repair contract the customer's items for repair are known as contract loan items.

c) Ordinary loan. This would apply where, say, the customer loans the seller items of special test-equipment or equipment into which the goods will ultimately interface, so that the seller may test the interface before delivery.

Data may be required as part of the detailed definition of the customer's requirements. As a simple example, a supplier of televisions would need to know certain technical information about the customer's power supply and the frequency of television transmissions in the customer's location. In practice, the customer should specify these parameters. If the contract is to supply a colour, stereo, teletext television and this data is not provided by the customer, the supplier may be able to get away with supplying a set that will not actually work. On the other hand, if the contract is to supply a colour, stereo, teletext television to operate in Benbecula, then the obligation on the seller is more specific and he will be foolish not to seek from the customer all the technical data necessary to perform the contract.

Whether the seller needs loan equipment or technical data from the buyer, there are some common ground-rules. In both cases:

a) The seller should advise the buyer as soon as possible of his needs.

b) The requirements should be specified as precisely as possible. Equipment should be defined in part number, specification, description or other unique identification.

c) The latest dates by which the equipment/data is required by the seller should be agreed.

d) Late, defective or incomplete provision of the equipment or data should be agreed as allowing the seller the right to a price change and/or change to the period of performance if these are consequent upon the late, defective, or incomplete provision.

It should be remembered that customer-furnished equipment will remain the property of the buyer whilst in the seller's possession, albeit that risk of loss or damage will be the seller's liability until the equipment has been returned to the buyer. The other key aspect, in so far as customer furnished data is concerned, is to ensure that the customer provides the seller with the right to copy and use the data for the purposes for the contract. If the data does not belong to the buyer he cannot give this right, and in which event the seller will expect the buyer to indemnify him against third-party claims of infringement of copyright or other intellectual property right.

Having specified the necessary customer-furnished equipment and data, having received it accordingly and checked it for damage or deficiencies, the supplier is then contractually bound to perform the contract. That is, if goods ultimately delivered to the customer are defective and rejected by the customer, the fact that customer-furnished equipment or data was utilised will not be a permissible excuse.

As a matter of good practice, requirements for customer-furnised equipment or data will be drawn up into a schedule and included as part of the contract documents to facilitate change and control. The schedule will be accompanied by a contract condition placing the obligation on the buyer to supply the equipment and data.

8

● ● ● ● ● ● ● **C** **H** **A** **P** **T** **E** **R** ● ●

The value
of ideas

① Introduction

In joining an established company a new recruit takes very many things for granted. The organisation, the canteen, the car park, a desk and chair and so on. He also takes for granted that the company has work in the factory and future orders yet to fulfil. Whatever the nature of the product, it was nevertheless at some much earlier time simply an idea. The idea grew and became developed into a product having a value in the market-place. The idea was the result of intellectual effort; the law will recognise the idea as being of the nature of property and will give the inventor the right to exploit it and prevent others from doing so. Thus we have the concept of there being legal rights in so-called intellectual property. The expression Intellectual Property Rights (IPR) is born.

National and international law surrounding IPR is both highly technical and demands detailed legal expertise. Large companies employ specialist legal departments to handle IPR matters. Smaller companies use skilled patent attornies and the one-man inventor is well advised to seek expert advice. What, then, is the relevance of this awful esoteric subject to engineers? The answer is several-fold:

a) It emphasises the monetary value of intellectual property. To a company the value of its ideas, inventions, processes, know-how and indeed its reputation is as material an asset as its plant and machinery.

b) It explains the various forms of intellectual property

and the rights which the law provides, such that understanding is enhanced.

 c) It explains the position of the employee.

 d) It explains how work is financed.

 e) To explain some straightforward practical measures to safeguard information.

 f) It points out the things to be considered when dealing with subcontractors.

② Commercial value

The commercial value of intellectual property can be realised in basically four ways:

 a) By selling a product which has been made by virtue of the intellectual property, which may be a unique design or a unique manufacturing process or both.

 b) By outright sale of the intellectual property. In selling an idea, just as in selling a product, the seller has transferred all legal rights to the buyer, who may do with the idea whatever he wishes.

 c) By permitting use by others of the intellectual property in return for some payment. This is known as licensing. Unlike a sale, the licensor retains ownership of and rights in the property and can exercise control over the use to which the licensee puts the property. Clearly the advantage to the licensee is that, by acquiring intellectual property, he acquires the ability to enter new technology or geographic markets. This may be an extremely attractive alternative to carrying out his own research and development, which may be prohibitively expensive and carries the risk of failure either to complete the development at all or to complete it in time to enter the particular market. On the other hand, the licensor has the opportunity to indirectly enter markets where hitherto he had been unable, perhaps because of export or import restrictions. Alternatively, the licensor may simply be seeking a continued income from designs which he may no longer wish himself to manufacture – for example, where in his own territory the design is 'Old technology' or where his own manufacturing facility has reached the end of its useful life.

The two parties to a licence might agree that the licensee will pay the licensor:

a) a royalty based on the value of sales made,

b) a fee not linked directly to the potential value to the licencee,

c) both.

The advantage in (*a*) is a potential continuing source of income for the licensor but with the risk that no sales will be made. A simple fee only, gives the licensee the risk of not achieving sufficient sales to recover the cost of the fee, but on the other hand, once he has broken even, he has no further payment to make to the licensor. A fee and royalties tend to be used where there is a transfer of technology.

d) By the enhancement of a product selling price arising from reputation/goodwill. The current trend in 'designer' clothes is an example of this. It has always been the case that certain 'names' will sell better and at a higher price than others, but more and more designers and manufacturers are realising the importance of this aspect of intellectual property.

③ Forms of intellectual property

3.1. Patents

The word 'patent' comes from the expression 'letter patent', which means an open letter of protection from the Crown. The principle is that, in return for disclosing the invention to the public, the Crown, through the medium of statute law, grants the inventor the exclusive right to exploit the invention for a specified period of time.

For an invention to be capable of being patented, three criteria must be satisfied:

a) The invention must be new.

b) There must be an 'inventive step'.

c) The invention must be capable of industrial exploitation.

The inventor has the right to seek patent protection and he must do this following a complex procedure through the UK Patent Office in London.

For a patent application to succeed the idea must not have

been previously made public.

Patent protection allows the inventor to take legal action against anybody else who is exploiting the invention even if that other person is not knowingly infringing the patent.

Once the patent is granted the invention is disclosed publicly. The Patent Office is a good source of information on the latest inventions. In making the invention known, any other person may take advantage of the general thinking behind the idea and it may help him to solve a design problem of his own. However, he may not exploit the actual invention which is the subject of the patent.

The Crown has the power for its own purposes to permit other than the patent holder to exploit the invention with impunity. In these circumstances the Crown will settle a licence fee directly with the patent holder.

3.2. Design rights

Whereas patent protection is concerned with the safeguarding of the underlying concept, design rights are concerned with the outward appearance.

A design may either be registered or unregistered.

If the creator decides to register his design he must do so following a procedure through the Designs Registry at the Patent Office. The Patent Office will make the details within the register available for public inspection but not the design itself. The design must be new and the protection afforded prevents others exploiting the design for a period of time. The design must be material to the proprietor. 'Must match' features are excluded.

A certain degree of protection is also provided for designs which are not registered. The registered design right provides that even an accidental infringement is nevertheless an infringement. For an unregistered design right to have been infringed there must have been copying. The unregistered design must be original and not commonplace. The protection does not extend to so-called 'must fit' and 'must match' features. This exclusion is specifically aimed at preventing suppliers claiming monopoly rights in the sale of spares or replacement parts.

The Crown has powers over designs similar to those it

possesses in connection with patents.

3.3. Copyright

Copyright covers the concrete expression of the invention in terms of drawings, diagrams, specifications, etc. Protection becomes available as soon as the material is created and infringement occurs only if there has been actual copying.

Computer software is considered to be copyright work and infringement includes activities such as electronically storing or displaying the software or adapting the software, including translation from one computer language to another.

Table 8.1. Summary of intellectual property rights

	Key features	Type of protection	Maximum period of protection
Patents	Covers technological inventions	Statutory	20 years
Registered designs	Covers appearance rather than concept	Statutory	25 years
Copyright	Covers expression of, rather than concept or appearance. Software is copyright	Statutory	50 years
Trade secrets	Unpatentable inventions. Non-technical information	Common Law	Indeterminable
Know how	Covers areas of expertise or skill	Common Law	Indeterminable
Trade marks or names	Covers symbols or a unique physical appearance	Statutory and Common Law	Indeterminable Indeterminable
Goodwill	Covers reputation	Common Law	Indeterminable
Unregistered designs	Covers appearance rather than concept	Statutory	10 years

Patents, design rights and copyright are the three areas of greatest interest. Other forms of intellectual property include trade secrets, trade marks and trading identities. A summary of the basic features is provided in Table 8.1. A number of free publications are available from the Patent Office which provide good readable additional material on this topic.

④ Employees' position

If an employee invents something as part of his job there are two sets of circumstances, in each of which the employer is entitled to claim ownership to the invention:

(1) If the employee makes his invention 'in the course of his normal duties' it belongs to his employer. A research worker inventing a new concept linked to his work could not claim the idea as his own. On the other hand a machine operator who out of the blue sees how his machine could be improved probably could.

(2) If, because of the employee's work and responsibilities,he has a 'special obligation' to further his employer's interests, anything he invents as a result of his job belongs to his employer. This is aimed particularly at senior staff – for example, a director of research and development might not be actively engaged in research but he would be expected to put his full skill and knowledge to his employer's service.

In all other circumstances the invention belongs to the employee and he is entitled to patent it and to receive any royalties arising from it. If the employer patents an invention made by one of his employees but which belongs in law to the employee, the employee is entitled to be named in the patent and enjoy its benefits.

Many companies put rules about employee's inventions into their employment contracts. Provided that the employee could reasonably have been expected to be aware of the rules when he entered into the contract, he is bound by them. However, the employer cannot use those rules to secure the right to an invention that has nothing to do with the employee's job. Any rule which seeks to deprive the employee of his rights in inventions belonging to him is void and unenforceable. Nor does the employer have the automatic right to an invention that, whilst partly related to the employee's job, does not arise directly from it and was worked upon by the employee in his own time, using his own materials and off his employer's premises.

If an employee invents something that, by law, becomes the property of his employer he may nevertheless be entitled to a 'fair share' of any profits. To succeed in his claim, the employee must show that his employer has derived 'outstanding benefit' from patenting the invention. The size and nature of the employer's business would be taken into account when deciding if the benefit has been outstanding. If the employer and the employee cannot agree on what is a fair share, the employee can apply to the Comptroller General of Patents, Designs and Trade Marks or to the Patent Court, a branch of the High Court, for an order requiring his employer to pay him compensation. The order requiring his employer to pay him compensation. The employee can lodge a claim for compensation at any time during the life of the patent, or within one year of the date of its expiry. Many companies run schemes that encourage and reward employees who make inventions of benefit or potential benefit to the company.

Copyright and design work also usually belong to the employer if created in the course or the employment.

⑤ Financing

The foregoing discussion in relation to patents and registered designs is in the general context of a firm employing people in the research, design and development of technologies and products and by that process inventing things that can become patented or registered as a registered design. The question then arises as to how this research, design and development (R+D) activity is paid for and whether the ultimate customers for the end product acquire any rights in the inventions and designs.

The sources of funding for R+D are as follows:

1) the company's own money;
2) customer's money;
3) grant from the government or from the EEC.

5.1. Company-funded

Where a company uses its own money for R+D the resultant product is called 'proprietary'. In practice the selling price for proprietary products will include an element to recover the cost of the R+D so that funds become available for further R+D of the same or other products. The selling price of proprietary products will also be set according to market forces. There is therefore considerable risk involved in this 'private venture' work and considerable market intelligence and commercial judgement needed to ensure an enterprise is a success. If the product needs an £8M R+D programme, costs £50 each to make, the world market is 100,000 in quantity and comparable products already sell at £150, the company would need to capture 80% of the market to 'break even' on the R+D investment and production costs. So to be successful the product must sell at a price and quantity that permits recovery of the R+D costs over a reasonable time-frame. The customer, although he is contributing to the R+D cost, acquires no rights in the design. He acquires total rights in the goods he has bought, i.e. he is free to use it, modify it, repair it however he wishes, but he secures no rights in the design. he is also, of course, free to sell or hire the product to others.

5.2. Customer-funded

If a customer has a particular requirement that cannot be met by existing off-the-shelf products and potential suppliers are unwilling or unable to undertake the special R+D needed, the customer may have no choice but to pay for the R+D himself. The supplier may be unwilling because the likely cost of the special R+D is too great for him to fund or perhaps because the perceived total market for the product is insufficient to recover the R+D costs. The supplier may be unable because he lacks the necessary numbers or combinations of skills or perhaps because special plant or equipment is necessary and again the perceived market is insufficient to warrant the expansion of resources or the capital investment to acquire special plant or equipment.In practice the customer and supplier may agree to share the cost of special R+D, capital investment, etc. Where the customer does directly fund R+D he will usually acquire rights in the design. These rights will have to be secured under the contract (i.e. rights do not automatically arise under statute or common law) and the extent of the rights will depend upon the proportion of the R+D cost contribution. The one exception to this rule is with the unregistered design right which belongs to the commissioner rather than the creator unless there is an agreement to the contrary.

5.3. Grant

The Government provides aid to industry in pursuit of three broad headings:

a) general R+D support, including certain specific project and collaborative ventures;

b) support for exports;

c) regional development.

The purpose is to promote job-creation, new technologies and new industry.

Under this type of grant, monies are available by industry for R+D purposes. The level of grant can vary enormously but typically might be between 25% and 50%. These grants, administered chiefly through the Department of Trade and Industry, are primarily concerned with achieving the political aims of job-creation and industry growth rather than meeting specific government requirements for goods or services. In the latter case HMG is very closely concerned with intellectual property rights wherever taxpayer's money is used for R+D. In the former case, although taxpayer's money is used, it is *de facto* used to promote the particular industry or company and therefore HMG's interest in intellectual property rights is generally limited to securing an undertaking from the company that an exploitation levy will be paid to HMG if the project is successful.

6 Safeguards

There may be several reasons for companies to exchange information. It may be in pursuance of a contractual obligation to another party or it may be for some mutual benefit, perhaps where the companies are teaming together to bid for new work. Where the information is commercially valuable, then no matter what the reason for the exchange of information, it is as well to establish a confidentiality agreement between the companies to protect the information and control its use.

Typically a confidentiality agreement will have the following features:

a) The word 'information' will be given a full definition to include oral, written, printed material, software and , perhaps by way of models, demonstrations, presentations.

b) Depending on the circumstances, the companies will decide whether they wish all information exchanged to be considered confidential or just specific material in which event it will need to be marked in some way. This decision in part will be based on the volume of information to be exchanged.

c) A statement will be included as to the proper use to which the information can be put.

d) Each side will agree to treat the other's information in like manner to the treatment of its own and only to divulge information on a need-to-know basis.

e) In some circumstances information cannot be considered as falling under the control of the confidentiality agreement. This includes:

1) Information already known to the other side.

2) Information which is available to the public at large.

3) Information received from a third party where the third party was entitled to divulge that information.

f) The receiving company acquires no intellectual property rights in information provided by the other company.

g) Set down will be the circumstances under which the agreement will come to an end. For example, this would be where the purpose of exchanging information had disappeared. Exchanging information to bid for a requirement which then is cancelled would be such a reason.

h) Whether or not the agreement is ended the companies may agree to carry on protecting exchanged information for several years. Ultimately any written material will be destroyed or returned to the originating side.

Confidentiality agreements are an essential device for the protection of commercially valuable information. In practice the parties to the agreement may appoint individuals to be solely responsible for controlling and policing the exchanges and protection of the information. The agreement will generally be signed between the companies. However, where a company sees that the information is particularly valuable and vital then the confidentiality aspects may be reinforced and emphasised by securing written confidentiality undertakings from individuals within the companies with whom information will be exchanged.

⑦ Subcontractors

It is most important in the area of intellectual property rights that sufficient rights are secured from, and obligations placed on, subcontractors and suppliers to enable the company to discharge its responsibilities under contract to its customers. For example, it could be financially disastrous for a company to accept a contract under which the customer must be given the free right to copy handbooks and manuals if these handbooks and manuals are to be provided by a subcontractor who has not accepted that the end-user has such an unlimited free right.

From the buyer's point of view it is prudent to seek a promise (referred to as indemnity) from the subcontractor that, if any third party should sue or claim against the buyer for infringement of intellectual property rights – arising from the use of intellectual property rights provided to him by the subcontractor – the subcontractor will settle any such claim or reimburse the buyer if it is he who settled the claim.

9

Inter-company relationships

① Introduction

In previous chapters the only relationship described between two companies has been that of straightforward buyer and seller. In this chapter is examined the somewhat different arrangement of prime and subcontractor and also of an even closer situation in which the companies are said to 'team' together.

The concept of prime contractorship arises from very large contracts where the customer knows what he wants, has the money but does not possess the necessary resources, technical skills or management skills to undertake the entire procurement by himself.

In placing one 'prime' contract he avoids the cost and effort of co-ordinating several contracts with various suppliers. On the other hand, he has put all his eggs into the one basket and takes the risk, notwithstanding any legal or contractual remedies available to him, that his chosen prime contractor is capable of taking on the responsibility of the full task.

Typically the responsibilities that a prime contractor may assume are:

a) Co-ordination of design.
b) Placing and administering of subcontracts.
c) Testing the system.
d) Planning and co-ordination.
e) Financial management.

② Co-ordination of design

For the customer the principle question is to what extent it is desirable or necessary for him to be involved in the design process when the major responsibility has devolved to the chosen prime contractor. On the one hand, provided the overall aim is well defined, the method of arriving at the overall aim may not matter. On the other hand, the prime contractor may wish to remain involved as an added safeguard in ensuring the technical integrity of the design and to ensure that, whilst the prime contractor is producing the lowest-cost design, it is not at the expense of other features such as ease of maintenance. If the customer wishes, for whatever reason, to remain involved it is necessary to consider the nature of the involvement. This may range from retaining the option to attend design-review meetings to requiring to review and approve all technical and other specifications of whatever level as and when they emerge. In this latter event the prime contractor will seek a contractual commitment from the customer that he will review and approve such documents within set periods of time. In the event of disagreement over technical issues the customer may wish to reserve to himself the final decision. This may lead the prime contractor to look for contractual protection against decisions of the customer with which he disagrees.

This puts the customer in an awkward position in so far as he cannot disagree in principle with his chosen prime contractor's need for protection against imposed technical direction, but conversely he is the customer and is entitled to get what he wants.

Anything more than superficial involvement by the customer tends to pervert the purpose of appointing a prime contractor in the first place. However, if the prime contractor is not 100% certain of his own technical competence – which may quite easily be the case where major subcontractors of specialist technical discipline are involved – then the possibility of the customer retaining detailed involvement and therefore some responsibility can seem quite attractive. The customer will be alert to this dilution of the prime-contractor's responsibility for design and will usually negotiate a contract condition to the effect that the prime-contractor's responsibility is not diminished by any act or approval of the customer unless specifically agreed in writing. This is not that unreasonable as, even where the customer reviews specifications, the prime contractor must have greater, more detailed technical knowledge and competence. Without a safeguard the prime contractor could offer for approval a specification which, with his detailed knowledge, he knows to be deficient and yet he knows that the customer does not have the ability to notice the deficiency. Therefore the prime contractor should be responsible for the deficiency unless he has specificially drawn it to the customer's attention and received written acceptance of it.

One of the advantages for the prime contractor in undertaking design co-ordination is in the acquisition of new technical skills and abilities through the employment of subcontractors of differing technical disciplines to his own. For example, a firm experienced in electronic telephone exchanges could acquire a level of expertise in radio transmissions if appointed as prime contractors for a telecommunications network in a remote location where transmission is to be by radio transmission rather than cable.

③ Subcontract

The customer may specify that certain subcontractors are to used for particular elements of the work or the use of subcontractors may be inevitable because of the nature of the work (for example, where the prime contractor does not possess all the necessary in-house skills) or where the size of the job is beyond the limitations of the prime contractor's own resources.

Where subcontractors are used – and for these purposes subcontractors are considered to be those suppliers having substantial amounts of the work and/or those with whom it is necessary to negotiate similar contract conditions as between the prime contractor and the customer – the customer saves himself considerable time and effort in passing responsibility for the negotiation, placing and subsequent administration of subcontracts. Commercially, the most significant point here is that the customer is giving the prime contractor contractual responsibility and liability for the elements of work which are to be subcontracted. In return for this the prime contractor is entitled to earn profit on the prices charged to him by the subcontractors. This is distinct from the situation in which the customer pays the prime contractor a management fee to negotiate and administer subcontracts which in practice are placed directly between the customer and the subcontractor. In these circumstances the prime contractor is acting as an agent only and is not a prime contractor as such.

The timing of events to permit effective subcontracting by the prime contractor is all-important. The basis of the subcontract will be as follows:

1) The inclusion of contract conditions which the customer requires to be included in all subcontracts. For example, an obligation to permit the customer access for inspection purposes.

2) The inclusion of contract conditions which are not required to be flowed down from the prime contract but which nevertheless it is prudent so to do; for example, cancellation-for-convenience. If this is in the prime contract but not in the subcontract and the customer cancelled the prime contract for convenience the prime contractor would be in breach of contract with his subcontractor if he failed to continue with the subcontract!

3) The inclusion of contract conditions which are designed to afford the prime contractor some protection against risks inherent in the prime contract. For example, liquidated damages. If the achievement of a major prime-contract milestone and payment is largely dependent upon the progress made by a subcontractor then it is commercially wise to include liquidated damages against the consequences of being prevented by the subcontractor's default from making achievements under the prime contract.

Thus the drafting and negotiation of subcontracts presumes the existence of a fully negotiated prime contract. Where this is not the case the prime contractor will have to judge what subcontract conditions are likely to be necessary in line with the foregoing. In some instances the negotiation of the prime contract and the subcontracts will be concurrent activities. The prime-contractor's aim in negotiation is to be in a minimum-risk situation by leaving as much risk as possible with the customer and by passing risk down to the subcontractors. All parties are intent on minimising their own risk, and as the man in the middle the prime contractor is possibly best placed to be the more successful in this regard. However, one thought which the prime contractor will have at the back of his mind is that on another day on another job he may find, himself as a subcontractor. 'Do as you would be done by' is a good maxim in this respect.

4 Testing the system

In accepting contractual responsibility for the system the prime contractor usually accepts an obligation to carry out testing of the complete system. Sub-elements can be tested at that level, but bringing the whole entity together and ensuring that it works as a system is where the prime contractor really starts to earn his money, and if the system does not work he will find out how successful he was in negotiating prime and subcontracts. This is so as the prime contractor's first instinct if the system fails test is to look for sub-system faults which may be subcontractor liability or a deficient requirement specification which may be the customer's liability.

5 Planning and co-ordination

Planning and co-ordinating the efforts of several hundred people or more, spread throughout many firms and possibly located in more than one country, is no mean task. All the firms involved have the same objectives – maximum profit, mimimum risk. The prime contractor's job is to manage and control activities to bring about the efficient and timely achievement of the tasks in hand.

6 Financial management

The degree of financial management which is required by, or visible to the customer is dictated by the type of contract. If the contract is fixed price at the outset, resulting from effective competition, then the amount and type of financial information to be passed to the customer is minimal.

In such a contract the prime contractor may provide no more than a forecast of invoicing, if indeed this requirement is not already satisfied by the inclusion of a stage-payment scheme with defined dates and invoice amounts. At the other extreme, if the contract is one under which the contractor and his subcontractors will be reimbursed their actual costs, then the financial information to be provided can be substantial and the need for financial management quite explicit. In a fixed-price contract financial management – a euphemism for cost suppression – is implicit. The existence of a fixed price pressurises all the contractors to progress quickly at minimum cost. For a cost reimbursement-type contract the prime contractor will be required to provide a highly detailed cost plan identifying estimated costs against contract-line items, major tasks and work packages, all of which must sit within the framework of a cohesive work-breakdown structure. Expenditure will be monitored on a quarterly or monthly basis by means of financial reports which must indicate actual spending against forecasts made in the cost plan. The prime contractor will operate the same arrangements between himself and his subcontractors, and the customer may have some visibility of subcontract cost plans and reports.

Prime contractor's risk

In subcontracting work – particularly if he is saddled with subcontractors nominated by the customer – the prime contractor is taking a major risk. The direct control and management of the work is removed from him and in the event that the subcontractor fails to perform there is no direct remedial action that he can take. Work is being undertaken geographically distanced from his own site and in any event the subcontractor's priorities are not likely to match his own. The prime contractor's performance under contract is at risk because it depends on the performance of the subcontractor.

⑧ Typical contract conditions

The prime contractor will seek to include in the subcontract the following:

- Virtually all of the prime-contract conditions, whether or not flow-down is manditory.
- Right of access to the subcontractor's premises for the purposes of inspection, quality assurance and general monitoring of the progress of work.
- The free right to copy, use and modify subcontractor intellectual property.
- The right not to pay the subcontractor until payment has been received from the customers.
- The right under contract to claim all damages, including special and consequential damages, from the subcontractor in the event of delay or default.
- The right to delivery penalties – but phrased carefully under the heading of liquidated damages to satisfy the requirements of English law – whether or not these apply in the prime contract.
- The right to withhold or recover payment in the event of delay.
- Extended warranty to cover any prime-contract warranty period.

The subcontractor may respond in the following terms:

- Only the prime-contract conditions which are mandatory flow-down should be included.
- Right of access is limited by commercial security considerations.
- Intellectual property can only be used for the purpose of the prime contract.
- Payment must be made within X days, whether or not the prime contractor has been paid.
- Liability for damages should be limited to a specific sum not exceeding the value of the subcontract.
- *Force majeure* protection against delays.
- A period of cure and consultation before any witholding or recovery of payment.
- Extra price for non-standard warranty.

Where the companies involved are of more or less equal standing and where from project to project the prime contractor/subcontractor, roles reverse, the rule of 'do as you would be done by' applies. A small firm who will only ever act in the subcontract role must recognise that he will have a tough time negotiating with his prime contractor and that he must carry a level of contract expertise that perhaps the size of his business does not otherwise warrant.

⑨ Teaming

There is a current trend for companies to 'team' together for some mutually beneficial aim, e.g. for the purpose of bidding for and winning a particular contract. The reason for the teaming is to bring together types and volumes of skills, plant, equipment and other resources which the team members individually do not possess. If the members agree to team on an exclusive basis, i.e. each agrees to work only within the team for the particular purpose and not to work with or for other companies, there can be a danger that the arrangement is illegal, being an anti-competitive practice which is against the public interest. If companies A, B and C all plan to bid for the contract and each ideally would wish to enhance its chances by association with the single source of some specialist skill – call it company D – then for any one of them to do so on an exclusive basis may be to effectively eliminate the other two from the competition. Depending on the circumstances, company D may wish to support one or all of them. There is an advantage in D not placing all its eggs in one basket. On the other hand, if the bid in question is of major proportions the cost of being involved with more than one customer may be prohibitive.

In teaming, the members agree that each remains an independent company. They agree what each may and may not do with respect to the purpose of the teaming but each remains a separate legal entity and each cannot act on behalf of the others or enter into commitments on behalf of the others, This degree of separation distinguishes the teaming agreement from other forms of collaboration. For example, the setting up of a consortium creates a separate legal entity to some extent independent of its constituent parts. A partnership, although not creating a separate legal entity, does permit each partner to act on behalf of the others. Members teaming in contract simply wish to join forces on an informal basis for the pursuit of a particular objective. Teaming agreements can be negotiated in a matter of days, other collaborative relationships may take much longer and are usually reserved for high-value and/or long-term projects.

Although the teaming arrangement is comparatively informal, the members may nevertheless wish to commit their purpose, undertandings and intentions to a signed written agreement so that each may appreciate the scope and limitations of the agreement. The negotiation of the details may be preceded by a 'heads of agreement' document setting out the main principle or principles.

The prospective parties to the teaming agreement may consider it necessary, before committing themselves to detailed discussion or before commencing any work, to establish the heads of agreement as a matter of good faith. The heads of agreement might simply describe an agreement 'to work together as independent contractors in the pursuit of contracts or orders in the civil telecommunications field and not to work with other companies in relation to the same market area'.

Once the broad agreement is in place, negotiations commence on the details of the teaming agreement. The main topics are:

1) A more comprehensive and precise defininition of the purpose or objective of the agreement.

2) A statment of the legal independence of the team members.

3) A statement of the legal relationship that will exist if the purpose of the teaming is achieved. For example, if the objective is to win a particular contract the parties must agree in whose name the tender will actually be submitted and therefore who will be the main contractor and who will be subcontractors in the event of success.

4) The allocation of work between the members in relation to the teaming phase and the succeeding phase.

5) The practices and procedures permitting the parties to actually work together. For example, the facilities each must make available at his site for the other to work.

6) The treatment of confidential information passed between the parties.

7) Limitations on publicity associated with the teaming.

8) The arrangements to be followed in the event of disputes.

9) Agreement on allocation of costs arising out of the teaming. Usually each member of the team is responsible for his own costs.

10) Events which cause termination of the teaming agreement. These may include mutual agreement to terminate or the disappearance of the objective for which the team was created. Occasionally the teaming agreement may continue to exist even after the objective is achieved. For example, a teaming agreement may continue to operate after the contract, if that was the objective when won. The disadvantage is that conflicts can arise unless there is a statement as to whether the teaming agreement or subsequent contractual relationship has precedence.

Where it is proposed to continue a teaming agreement it is essential that thought is given to the practicalities of the operation of the team in so far as they need to be formalised in a written agreement. Given also that there could be practical difficulties, it is important to open the formal discussions as early as possible to ensure that any potential difficulties are identified quickly and dealt with in appropriate fashion. A particular example of this involved the bidding for a contract, wherein there would be three batches of deliveries, each to different countries, A, B and C. The second and third batches were options only, and could be taken up by either the Procurement Authority in A or by the Procurement Authority in B and C as appropriate. So company X could expect three orders, all from A or one order each from A, B and C. So far so good. Company X felt that its chances of winning would be enhanced by teaming with a company Y, from country B. The teaming envisaged would have X taking the primary role for orders from A and C, and Y taking the primary role in orders from B. This may not have been too complicated except that at the time of tendering it was not known whether the order for B would be placed by A or B. Thus the permutations of potential contractual arrangements were as shown in Table 9.1.

Table 9.1.

	Order for A	Order for B				Order for C	
Permutation	a	b	c	d	e	f	g
Procurement authority	A	A	A	B	B	A	C
Prime contractor	X	X	Y	X	Y	X	X
Subcontractor	Y	Y	X	Y	X	Y	Y

Assuming that it would have been possible to agree a set of words to describe all this in the short time that there was available, especially since X and Y spoke different languages, crucial difficulties were:

1) Since at the outset the tender had to be submitted by X to A then for Y to include in the tender permutation c or e would have required the teaming agreement to give X powers to make an offer on behalf of Y – authority to act on behalf of another is outside the scope of a simple teaming agreement.

2) For X to offer A permutation c presumed that A would have been willing to contract with Y, which was not necessarily the case.

In practice, company Y only wanted to team in permutation c, and e could be offered, but in the time avaliable Y could not get authority to let X make an offer on its behalf, and X was not wholly keen on the idea anyway. In the circumstances it was decided to drop the proposed teaming and X tendered by itself. The important point is that the contracts people were involved early enough to expose these problems before both companies had become embroiled in the work of jointly preparing a tender that probably would never have been authorised for release.

10

Negotiation

1 Introduction

The *OED* defines the word 'negotiate' as 'confer with a view to compromise'. This is an excellent definition because it encapsulates the concept of compromise, the idea that both sides must be prepared to yield on some points if agreement is to be reached.

In business there are many things requiring negotiation and the engineer may typically be involved in discussions on the detail of technical specifications, programmes, manufacturing plans and the like. There are occasions when the engineer is needed to support commercial negotiations, and where this is the case his role, and the manner in which it is played, is most important.

There have been written very many books and manuals on the art and skills of negotiation. The reader is commended to these invaluable works as they provide instruction in many aspects, such as:

a) venue, timing, etc.,
b) positional versus principled argument,
c) role-playing,
d) hard man/soft man,
e) humour,
f) stonewalling,
g) the 'walkout',
h) individual techniques,
i) aggression.

As well as illustrating the approach that one's own side may adopt, background material such as this may provide an insight to the opposition's plans and tactics.

The purpose of this chapter, however, is not to repeat what can be read elsewhere in specialist volumes but to draw out key points which experience shows are fundamental and in which the engineer can best make his contribution to the overall commercial success.

② Purpose

Before describing these points it is pertinent to focus on the purpose and importance of the commercial negotiation. The negotiation itself might be related to:

a) A pre-contract issue.

b) The process of agreeing the contract.

c) A difference of interpretation during contract performance.

d) A failure by the other party affecting contract performance.

e) A third-party act affecting contract performance.

f) An event affecting the contract but not previously contemplated and therefore not legislated for within the contract.

g) A dispute over completion or acceptance of the work.

h) A post-contract claim.

Whatever the issue generating the need for a negotiation the seller potentially has at stake:

a) profit,

b) cash,

c) reputation,

d) nugatory work,

e) diversion of effort,

f) delay,

and the buyer potentially has at stake:

a) profit,

b) cash,

c) reputation,

d) nugatory work,

e) diversion of effort,

f) delay.

Thus, and not surprisingly, the two parties each may have as much to lose or win as the other. Cynically put, one side's profit is the other's loss – although this, as will be seen, is an over-simplistic view of what transpires in negotiation. Nevertheless the point is well made that each party is there faced in the extreme with success or disaster. That is to say, when all the normal processes by which progress is achieved and agreements made have been exhausted, the parties must meet face-to-face to negotiate. Letters, routine meetings, minutes, phone calls, have all failed and it is down to the negotiators to thrash out a compromise. Indeed, it is a well-established rule of thumb and good practice to follow that, if a single round of correspondence has failed to resolve a problem once it has become an issue, then a negotiation should be the next step.

Further rounds of formal correspondence are likely to cause only entrenchment of the respective positions and thus inhibit a satisfactory agreement as both sides feel bound not to yield principles that the written correspondence record as being immutable.

③ Prior events

The engineer should note that almost invariably negotiations are preceded by events that are recorded or by propositions that are argued in writing. Care should always be taken in what is put in writing, and in other-than-routine matters advice should be sought on what to say and how to say it. Many an engineer has been caught out by unintentionally or unconsciously giving a commitment or yielding a principle that subsequently proves to be vital.

④ Serious matters

Whilst care must be taken, these comments are not intended to caution the engineer not to put things in writing to the other party. Indeed, in any commercial negotiation – and in extreme in front of a court of law – the volume and accuracy of documentary material is important and may be crucial. In general, factual information is most important (dates of meetings, events, records of phone calls) but opinion and speculation should be avoided. In particular, the following should not be discussed without considered prior thought:

    a) Costs/price.

    b) Cash flow/payment arrangements.

    c) Programme/delivery.

    d) Liability for delay, mistakes, accidents.

    e) Liability for poor technical performance, reliability.

    f) Ownership of design rights.

These, after all, are matters for the contract or tender and are not to be treated lightly.

In summary, it can be said that:

a) The negotiation is a crucial milestone and one upon which the company may stand to win or lose its shirt.

b) Most negotiations are preceded by some degree of correspondence or documentary material which is, or becomes, germane to the negotiation.

The commercial engineer will continuously bear in mind the sensitivity required in written material and will practice the skills in his support to the actual negotiation.

And so to the key points:

5 Preparation

Almost all successful negotiations are based upon thorough preparation. Too often the date for the negotiation is fixed as the first available date in people's diaries. If it is the first available date then impliedly the days leading up to the agreed date are already planned to be occupied by other events – holidays, meetings, etc. In those circumstances preparation for the negotiation has to be squeezed in amongst many other things. Similarly, if the preparatory meeting is left to the last minute then it may be too late to gather information or consult other people where these needs are only identified in the pre-meeting discussions.

Rule: Allow sufficient time.

It is the responsibility of the intended leader of the negotiation to arrange the pre-meeting. It is the responsibility of the engineer to make the time available to support and attend the pre-meeting.

The pre-meeting must be seen by all participants as absolutely crucial. It is the meeting at which the plan for the negotiation will be thrashed out and clearly it is madness for participants to the negotiations not to meet beforehand. It is wrong to think that the participants have all necessary information in their heads, can simply meet on the day and that a successful negotiation will ensue.

Rule: The preparatory meeting is crucial.

The primary objective of the pre-meeting is to establish the negotiation plan. However, equally vital is the gathering and assimilation of information. It is in this area that the engineer can make his greatest contribution to the preparation stage. The lead negotiator must have to hand all related information so that he can analyse what is useful to him, what is useful to the other side and what can be discarded. Notice that for this purpose he needs *related* information rather than just *relevant* information. That is to say, irrelevant information may have its place in the negotiation by way of red herrings, diversionary tactics, time-wasting tangents. Although it sounds senseless, the hardest part of the process is in thinking what is related to the issue in question. At the pre-meeting the lead negotiator will already have a number of questions for the other participants to answer or investigate and report upon. However, the engineer may have been involved with the other side for a considerable period of time and has been involved in many meetings, reports, discussions, presentations, demonstrations. Amongst the enormous mass of information that this represents the engineer must recollect and volunteer almost everything that comes to mind. He should never conceal anything that he or the company has said or done or anything that the other side has said or done which weakens his own side's position. If weaknesses are identified the negotiation plan can accommodate counter-measures to be deployed if and when the other side attacks the weaknesses. The pre-meeting is also the opportune time to identify the people and personalities on the opposition team.

Rule: Maximise information.

In conceptual terms, the objective of the pre-meeting is to create the negotiation plan, but if we can return to the definition of the 'confer with a view to compromise' then the essence of the plan must be a clear view of one's own ultimate objectives, a good view of the other side's likely objectives and in *both* cases a view of how the objectives might be compromised in order that agreement can be reached.

Rule: Identify both side's objectives.

In considering the opposition's objectives there is a subtle distinction to be drawn between his objectives and what is actually important to him. For example, if he feels considerably aggrieved, one objective may be to extract a price reduction, i.e. some financial retribution for his hurt. However, what may be of far greater importance is improved delivery or different bells and whistles. Again, background intelligence on the opposition's real needs may very well be something that the engineer may have, and this key information must be fed to the lead negotiator.

Rule: Identify the opposition's real needs.

A further facet of the opposition' s likely situation is that there may be areas on which he really cannot compromise. On occasions these may appear to be trivial and on the face of it represent points to be easily won. However, there may be valid reasons for the other side being unable to give in and it is most helpful to know what these issues are. Once known, then depending upon the progress of the negotiation and upon the negotiation plan, these points may be yielded to help-him-so-he-will-help-you, or at the other extreme they may be used as a means to contrive a breakdown in the negotiations – all through the other side's intransigence.

Rule: Identify what is difficult for the opposition to yield.

These last few points could more generally be described as 'identify strengths and weaknesses'. However, the aim has been to go beyond such a simple statement and indicate the type of information which will help in the development of the negotiation plan.

6 The negotiation plan

Montgomery – or it may have been Rommel, or perhaps both of them – said 'planning is everthing but nothing ever goes according to plan'. This wonderful statement encapsulates the principle that no matter how thorough the preparation, how good the rehearsal, how comprehensive the plan, there will inevitably be unexpected and unpredictable events, questions, issues and difficulties to deal with. Nevertheless, the existence of a plan with its goals and objectives gives a stable framework from which to diverge and also a framework to aim to return to once the unexpected has been dealt with.

Rule: There must be a plan.

The plan itself must enshrine some of the work done in the preparation phrase. Clearly the plan must centre on the

		Opposition's position		
		Impossible	*Difficult*	*Easy*
Objectives	*Must have*			
	Nice to have			
	Bonuses			

objectives of the negotiation and it can be useful to develop a matrix as shown above.

If there are objectives in the 'must have' but 'impossible to attain' segment then the purpose of the negotiation must be re-examined and if necessary redefined. Possibly the purpose devolves, on the other hand, to simply maintaining (or opening) dialogue, or, on the other, to commencing a series of negotiations.

In technical matters the engineer has a major role to play in both helping to define the objectives and classifying them between the nine segments of the matrix.

Rule: Categorise your objectives.

Similarly, it is essential that if the opposition's objectives and real needs have been identified then a matrix can also be drawn (see below).

Once again, topics in the 'opposition must have' but 'we must not agree' must cause a reassessment of the purpose of the negotiation.

		Your position		
		Must not agree	*Yield under pressure*	*Throwaways*
Opposition objectives	*Must have*			
	Nice to have			
	Bonuses			

Rule: Categorise his objectives.

It is the task of the engineer to ensure that technical objectives – yours and his – are viable and to help the lead negotiator understand the true position. For example, if you tell the lead negotiator that the opposition's delivery time must be halved and yet you know that component lead times make this impossible you are inviting trouble in the negotiation – particularly if your side is made to look foolish for apparently not knowing a common fact. Similarly, if an objective of the other side is that the power output of your product must be increased by 15% then do not let the lead negotiator believe he can agree 10% when the product is already operating beyond its expected performance.

Rule: Be realistic.

On this question of advice to the lead negotiator, it is the responsibility of the engineer, and indeed any other individuals involved in the preparation of or involved in the negotiation, to brief the lead negotiator thoroughly and honestly. For example, in connection with cost or time estimates the engineer should not keep a 'bit of contingency' to himself. The lead negotiator must know the complete and real picture if he is to successfully prosecute the deal.

Rule: Thoroughly brief the lead negotiator.

Having covered the objective the plan must have an intended structure for the negotiation. If the discussions proceed to your plan then the chances of success are improved. The structure should include the following topics:

a) Opening position.
b) Order of play.
c) Manner of tabling issues.
d) Timing of offers/counter offers.
e) Information to be tabled.

and, where appropriate, the timing and mechanics for breaking for lunch, time-outs, etc.

Rule: Structure the negotiation.

Most importantly, the negotiating team must have clearly defined roles and contributions to make. Each member must be fully briefed on his own specialist subject and on the generality of the negotiation plan.

Rule: Allocate specific functions to each member of the negotiation team.

It should be a taken-as-read rule of all negotiations that the team will follow the leader.

7 Follow the leader

Experience shows that in the vast majority of negotiations the actual process of negotiation occurs between just two people – one on each side – regardless of the respective team sizes. This is hardly surprising since the negotiation may be likened to a game of chess. It would be ludicrous to think of a chess game being played equally well be a team as by a single person. The single person controls the tactics within an overall strategy, sacrifices pieces or positions and captures pieces or territory as part of an overall game plan. The golden rule must be that the negotiation is the responsibility of the leader and all other participants are there to support him.

Rule: The leader is responsible.

To effectively support the leader the engineer must develop listening skills so that he is sensitive to the line being pursued by the leader, to its probable purpose and such that he is alert and prepared to lend supporting argument if invited to do so by the leader.

Rule: Listen well and be sensitive.

One of the worst things that can happen to the leader is for one of his team accidentally or, even worse, intentionally, in the heat of the moment, to usurp the leader's role. If the problem is bad then the seriousness is increased tenfold if the usurper or indeed a supporter deviates from the agreed negotiation plan. At best this will cause confusion and at worst is a recipe for disaster.

Rule: Stick to the plan.

However, as Montgomery said, nothing ever goes according to plan. Diversions from the plan must be orchestrated by the leader and the team, if they have been listening well and have been sensitive to the general trend in the negotiation, will see how and to what extent the plan is being abandoned. The leader will be conscious of the need to let the team know his thinking and will find a way of bringing everyone up to speed, even if it means calling a time-out.

Rule: Recognise when the leader is deviating from the plan.

A further action to be avoided is the tabling openly at the meeting of 'new' information. It is not unnatural as the discussion proceeds for the engineer to recall additional data, to see a new slant on certain things or indeed to have been given further information after the pre-meeting but before the negotiation. Whatever the reason, the worst thing he can do is to table it, leaving the leader with the awful prospect of trying to recover a self-inflicted wound to the foot. If there is new information to consider it should be communicated to the leader. Again, it is preferrable to have a time-out rather than potentially destroy the painstakingly achieved progress through ill-timed presentation of extra information.

Rule: Don't throw in new information.

Regardless of the leader/supporter functionality, the team is nevertheless there as a team representing the company. It sould be seen by the opposition as being united and of a common view. It destroys the credibility of the entire enterprise if the team appears to be divided, or not equally briefed or, in the very worst extreme, antagonistic to one another. For example, if the engineer should say to the leader 'actually I disagree with you there' then the negotiation will almost certainly plummet. It is an unforgivable mistake which the opposition should ruthlessly exploit. There may very well be disagreement but it should not be made public, and if the engineer believes the leader has got it seriously wrong he must find a discreet (but rapid) means of letting the leader know.

Rule: Don't drop the team in it.

Having got to the point where the negotiation is in progress, everybody is supporting the leader and generally sticking to the plan, there are nevertheless a variety of other fundamentals which need to be taken on board if the engineer is to maximise his contribution.

⑧ Other fundamentals

In many cases the commercial negotiator shows a certain reluctance to take along an engineer in support of the negotiation. This reluctance flows from a great fear that the engineer will hinder, not help, through his not-unnatural wish to answer questions. If there are technical issues to resolve then the presence of the engineer may be unavoidable. From the point of view of the lead negotiator there is a great danger that the engineer will then rush off in his eagerness to display his deep understanding and knowledge of the subject to answer all the wrong questions from the other side. All of us have the basic intellectual desire to demonstrate that we can answer the question. In a negotiation the other side will be seeking to put questions on areas in which they perceive your position to be weak. To answer such questions fully and frankly is to pull the proverbial rug from beneath one's feet. Far better, within reason, to feign ignorance, promise to check later or find some other way to avoid giving a direct answer.

Rule: Avoid answering awkward questions.

Even worse than answering awkward questions is answering questions in areas where you have no expert knowledge. Although the aim of this book is to broaden an engineer's knowledge into wider commercial fields he should still, by and large, leave the answering of the commercial questions to the lead negotiator. The leader has enough problems with which to deal without having to keep his own side under control and without having to interrupt his own people to prevent them going off into the wrong subject.

Rule: Don't stray into areas in which you are not the expert.

Another facet of the answering of questions – where, of course, it is in your interest do so to – is to be economical with the truth. From a legal viewpoint there are significant risks in lying your way out, as you and the company may be guilty of fraud, attempted fraud, or misrepresentation. Economy with the truth is not free of these risks but the general objective must be to answer questions with the minimum of information.

Rule: Be economical in answering questions.

There is good information and bad information when it comes to the actual negotiation. Beforehand, all information is good, even if in itself it is bad news, but once at the meeting one of the worst statements the engineer can make is one that begins with the words 'I don't know if this helps or not', almost certainly, if he does not know, the chances are that whatever he is going to say will help the other side. In any event he is giving his own side no more time than the opposition to consider the information and its implications.

Rule: Don't volunteer information.

One of the golden rules in any negotiation is not to lose sight of the objectives. It can be all too easy to get carried away and perhaps submerged in vast volumes of information, intellectual debates, slanging matches or whatever, but the light at the end of the tunnel is the objective of having the meeting in the first place and in sight it should be pursued relentlessly.

Rule: Don't lose sight of the objective(s).

A feature of the negotiation that often comes as something of a surprise to the engineer is the apparent attitude of the opposition and the atmosphere in which it is conducted. Whatever the subject of the negotiation, it is likely to have cropped up as an issue following a considerable period of friendly relations. For example, where a contract has been running smoothly for some time, the interface between the two parties will have been amicable. Routine progress meetings and reviews will have passed off with the two sides congratulating themselves on their close co-operation and the friendly manner in which the business is conducted. Suddenly there is an issue to resolve which normal activities and correspondence have failed to settle. A negotiation is necessary and the other side appears in a far more aggressive, difficult and formal manner than previously. This of course is not really surprising and indeed one' s own side will be in a similar mode. It should therefore be seen as a normal part of negotiation and, whilst excessive aggression or rigidness is rarely a recipe for success, it should not be found off-putting.

Rule: Don't forget it's tough.

Even though the negotiation should be expected to be tough it is a very stimulating and potentially rewarding activity. By and large the best negotiators are those who not only have the technical and personal skills to carry it off successfully but also enjoy it for its own sake. Similarly, the engineer should see his participation in the same light and hopefully expect not only to make his contribution but also to take some degree of personal satisfaction from it.

Rule: Don't forget to have fun.

The final fundamental is the concept of the good deal, i.e. a deal which both sides find satisfactory. It is unusual to be negotiating with another party where there has been no previous business with them nor any likelihood of further transactions in the future. Thus, if an overall balance of good relations is to be maintained, both sides must leave the table content, more or less, with their half of the bargain. To return again to the definition: the purpose is for both sides to compromise. The negotiation must not be seen as a pitched battle from which there must emerge a clear victor and a clear loser.

It is the job of the lead neogotiator as the negotiation draws to a conclusion to decide if overall, he is being offered an acceptable compromise. Amongst many other things, in making his decision he will take into account the consequences of not making an agreement on the day. The consequences may include delayed payment, late delivery, lost profit, and indeed he must also consider the chances of improving upon the offer if he decides to wait for a further negotiation at some later date. The engineer should be aware that these deliberations are running through the mind of the leader and he should support the decision whatever it may be. These considerations applying equally to both sides.

Rule: *Observe the 'good deal' principles.*

Finally, it should not be forgotten that the other side will have good points, and in the extreme the opposition may actually be in the right and you in the wrong. This is no reason, though, to go agreeing with him. Similarly, the engineer must resist the temptation to support the opposition's case, which at worst can have the engineer actually appearing to change sides.

Rule: *Don't argue the other side's case for him.*

⑨ A summary of the rules

The various issues and practicalities that the engineer should bear in mind are summarised as follows:

Preparation

Allow sufficient time.

The preparatory meeting is crucial.

Maximise information.

Identify both side's objectives.

Identify the opposition's real needs.

Identify what is difficult for the opposition.
Negotiation plan
There must be a plan.
Categorise your objectives.
Categorise his objectives.
Be realistic.
Brief the leader.
Structure the negotiation.
Allocate specific functions.
Follow the leader
The leader is responsible.
Listen well.
Stick to the plan.
Recognise deviations.
Don't throw in new information.
Don't drop the team in it.
Other fundamentals
Avoid answering awkward questions.
Don't stray.
Be economical in answers.
Don't volunteer information.
Have fun.
Remember the 'good deal'.
Don't argue the other side's case.

⑩ Examples

To illustrate the operation of some of the these principles there follows some actual examples where the commercial negotiation was scuppered, jeopardised or hindered by well-intentioned engineers failing to observe the self-preservation rules.

Example 1

The customer wished to place a contract for the supply, installation and commissioning of electronic equipment. The scope of the work was not fully defined and some actual development work would also be necessary. The key features of the customer's needs were that the requirement could only be satisfied by a single supplier and that delivery on time was absolutely crucial.

The supplier was in a very good position to negotiate a favourable contract. He decided that in the particular circumstances his optimum method for exploiting the opportunity in terms of increased profitability was to negotiate a bonus scheme for timely delivery rather than, for example, attempting to charge high prices. The negotiation plan was thrashed out and centred on two key principles:

1) The supplier would formally table a programme of work and delivery dates which clearly showed that under a conventional contract the customer's vital date would be missed.

2) Outline details would be proposed for a cash bonus scheme that would indicate a high probability of meeting the vital date.

The supplier's negotiation team would be lead by the commercial manager with the project manager in support. A preparatory meeting was held and the role of the project manager was defined as being to provide technical arguments and reasons to explain why the work was difficult, why delivery could not be brought forward under a conventional contract, and the special measures that could be taken if a bonus scheme were to apply. Crucial to the success of the strategy would be resistence to pressure from the customer to improve delivery with no bonus scheme.

The negotiation meeting commenced and, exactly as predicted in the pre-meeting, the customer's angle was:

a) The importance of timely delivery.

b) His wish to place a contract quickly.

c) His difficulty in acquiring authority to include a bonus scheme.

The supplier stuck to his line until in desperation the customer said: 'Look, if we place a conventional contract with you, can you advance the delivery plan you've offered?' The supplier's project manager leapt in with words equivalent to 'of course'. The supplier had thereby pulled the rug from beneath his own feet and his position, so carefully developed, collapsed.

It is not difficult to see the mistakes of the supplier's project manager. These were:

a) He did not stick to the plan.

b) He did not follow the leader.

c) He gave a straight answer to an awkward question.

Additionally he yielded to the emotional 'cry-for-help' pressure, forgetting the 'it's tough' rule.

Example 2

The company had arranged a meeting with three other companies to negotiate the essential features of a teaming agreement. The features were the scope of work, volume of effort and applicable prices – these three variables being interdependent. The nature of the overall job was such that each of the four companies would wish to secure a wide scope of work, a sizeable volume of effort and, not surprisingly, high prices. The lead company needed to settle a compromise of these 12 variables between the four organisations within an overall ceiling value.

A preparatory meeting was held although the project manager was unavailable. The agreed tactic was to concentrate on the arguments which would be put forward by the other three, each in defending its own position, and to leave the company's own contribution to last in the probability that it would go through almost on the nod.

The negotiation started well, with close adherence to the plan. However, the company could not escape having its own contribution interrogated. Choosing his words carefully the lead negotiator asked his project manager to describe the technical nature of the work (as opposed to asking him to defend the level of effort proposed). The project manager replied by saying that the leader was probably not yet aware of some paper in the system that conveyed a large reduction in the company's planned amount of effort. Whilst this may have been true, it nevertheless, undermined the company's position at a stroke. In this example the mistakes made were:

a) The project manager did not attend the pre-meeting.

b) Therefore there was no chance at all of everybody sticking to the plan.

c) The project manager did not listen properly.

d) The project manager threw in new information.

Example 3

The supplier and customer had been involved in a long

running argument over whether the supplier's proposed design met the requirements of the contract. The customer had steadfastly maintained that the design was not sufficient and that the supplier must change the design at his own cost. The supplier was in a diametrically opposed position. It had proved impossible to resolve the issue on purely technical grounds as these aspects were extremely complex and the issue hinged on differing interpretation and differing expert opinion on the two sides. Whilst arbitration and recourse to the courts was of course open to the two sides, it was in neither's interest to follow such a course of action. To do so would have meant the customer tolerating unacceptable delays in delivery – there being no practicable alternative method available to him. The supplier would have been starved of cash-flow for an extensive period of time.

In an attempt to move forward, the supplier and the customer had agreed without prejudice to each's position that the supplier should put forward a change to the design and a price quotation for the additional work. It was quickly agreed that the change was desirable and that it would correct the perceived weakness in the design. This of course did not settle the question of who should pay.

The customer called the supplier to a meeting at the customer's premises, which were geographically considerably distant from those of the supplier. The commercial manager and project manager had a pre-meeting at which it was agreed that if the customer was not preparing to alter his stance he would have communicated the fact by telephone or in writing. Thus it was a good sign that the customer sought a face-to-face meeting, particulary in view of the distance involved, as at the very least the customer was known not to waste people's time capriciously. The plan for the meeting was that a compromise on possible splitting of the costs would be acceptable, albeit that it would be a hard slog fought for inch by inch.

The prediction about the meeting proved to be accurate and the customer repeatedly came down heavily with his much-stated view that he carried no liability whatsoever for additional costs. Nevertheless, slowly but surely progress was made towards a compromise settlement until, in nothing short of exasperation, the supplier's project manager blurted out the question, 'if you're not liable, then you won't pay and why have you called us to the this meeting'. Although not catastrophic, this put at risk the progress so far made and almost invited the customer to bring the shutters down and entertain no further discussion. the rules ignored were:

    a) Stick to the plan.

    b) Follow the leader.

    c) Don't forget it's tough.

Example 4

The supplier had arranged a meeting with the customer to discuss a potential order for a range of products which the customer had bought previously from the supplier. The nature of the work was relatively involved and included elements to be supplied by several major subcontractors. There were many issues to be discussed, including the question of where certain risks were to be carried. The supplier's aim was to establish the principle that as, between himself and the customer, he, the supplier, carried the risk. This was an important point as the price to be agreed would be linked to, amongst other things, the level of risk inherent in the job.

At the pre-meeting it was agreed that the project engineer would support the commercial manager and answer only those questions which the commercial manager put directly to him.

At the meeting with the customer the supplier put forward the various relevant arguments to sustain the principle that he was proposing. As a natural part of the process the commercial manager gave examples of how in practice a real problem would fall to his account since he carried the risk. At the worst possible moment the project engineer interrupted to say that in the examples given, that was not how the problems had been dealt with in previous contracts. The interruption was in part made in frustration, through his not understanding the significance of the apparently irrelevant principles. This statement caused the negotiation to take five steps backwards from the point of view of the supplier. Once again some basic rules had been overlooked in the heat of the moment:

    a) Stick to the plan.

    b) Don't drop the team in it.

c) Don't stray.

d) Follow the leader.

Of course, the engineer had completely lost sight of the objectives.

These examples are based upon actual events and are not exaggerated for effect. Apparently trivial or helpful comments, no matter how well intentioned, can destroy a carefully constructed negotiation. From the lead negotiator's point of view these major or even minor disasters present immediate problems:

a) To recover the ground lost with the other side.

b) To explain away his own side's comments.

c) To avoid an argument developing on his own side.

d) To maintain his side's professionalism.

(11) Chain of events

To return to a theme from the opening pargraphs of this chapter, it is usually the case that a negotiation is preceded by a chain of events, meetings and correspondence, most of which at the time may seem to be routine. However, since earlier events are bound to feature in the negotiation, everybody having contact with the other side must keep at the front of his mind the potential benefit or potential damage that specific statements, promises and actions may have in the future. In particular, it is as well to remember the following:

a) If the other side has said or done something of value to me, I must get him to confirm it in writing, failing which I will write to him to confirm.

b) If I have said or done something of value to me, I will write to the other side to confirm it and seek his written acknowledgement.

The value of documentary evidence – formal correspondence, minutes of meetings, notes of telephone calls, etc – cannot be overstated.

Events leading up to the negotiation can in themselves have such a material impact on success or otherwise. For example, within a proposal that had been sent to the customer was a particular assumption forming a key part of the offer, albeit described in fairly innocuous terms. The customer picked it up and queried it on the engineering interface with the supplier. The engineer formed his own view that it was trivial and lead the customer to believe that it was of no consequence at all. As a deliberate, well-thought-out tactic, this possibly could have a legitimate ploy. In the actual circumstances the engineer allowed the customer to believe that the point was not seriously made. When the negotiation meeting commenced the supplier then had an uphill struggle to convince the customer that he was indeed serious. The moral of this story is to be found in a further general rule describing the engineer's role in a commercial negotiation:

> *The engineer must work in close contact with his commerical negotiator in preparing the ground with the opposition.*

This truism applies whether or not the engineer is actually to be present at the negotiation.

⑫ Event-planning

Using the analogy of the chess game, the negotiation itself may be likened to the 'end game'. Extending the analogy to the levels at which chess is played, two distinct variations can be seen:

Beginners level. In which the two players move their pieces almost at random until by chance an opportunity for checkmate arises and the end-game is only then conceived.

Advanced level. In which the end-game is planned from the outset and each move by the opponent is countered by moves that aim to restore progress towards the end-game.

In business the equivalent to the 'beginners level' is a situation in which events are following a normal and natural course, a crisis or problem then emerges and a negotiation then takes place to effect resolution. The equivalent to the 'advanced level' is a situation in which it is estimated from the outset that a negotiation will be necessary before agreement is reached, and thus all events over which control can be exercised must be pre-planned to achieve progress towards the desired end. Just as in advanced chess, the side which can then see the greatest number of moves ahead and predict the opposition's counter-moves is the more likely to succeed.

Where the company is involved in an 'advanced' game the importance of the engineer liaising closely with his commercial department cannot be overstressed.

⑬ A final rule

The final rule is one which applies to all participants in the negotiation, whether of engineering, commercial or other background. Put very simply, the rule is

Think Before You Speak.

Or put more bluntly – *engage brain before mouth*.

The knack, of course, is to follow the politicians' habit of talking without really saying anthing. Whilst many people would say that politicians take this to the extreme, the principle is nevertheless sound. To sit in silence whilst thinking about the answer is almost as bad as giving an ill-considered reply. Therefore to talk around the subject or off it entirely allows the prospect of, at best, taking the discussion along some alternative route which the better suits your purpose or, at worst, allows thinking time before giving a substantive response.

11

○ ○ ○ ○ ○ ○ C H A P T E R ○ ○

Contributing
to success

① Introduction

The purpose of this book has been to convey some of the principles and practices that describe the commercial operation of business, from the construction of a company and the legal operation of its contracts to the paramount objectives of profit and growth. This chapter will draw together the key commercial factors that should be consciously borne in mind and to which a contribution can be made.

② Organisation

In many engineering firms the number of engineers naturally and rightly swamps the number of commercial people. A typical distribution of numbers by discipline is shown in Fig 11.1. The disadvantage for most engineers in these circumstances is that they are not able to see the broad picture, dealing only with a relatively small proportion of the company's products, projects, contracts or customers. Thus it is necessary for engineers to input to and consult with the functions that are able to see across the spectrum of the business operation. Indeed it is important to understand what each discipline within the company is responsible for and how it operates. The Finance department may be responsible for project, statutory and management accounting. Marketing and Sales may embrace business development. Commercial may include legal and purchasing functions. It is necessary to understand the chains of command,

Figure 11.1: Comparative sizes of functional departments

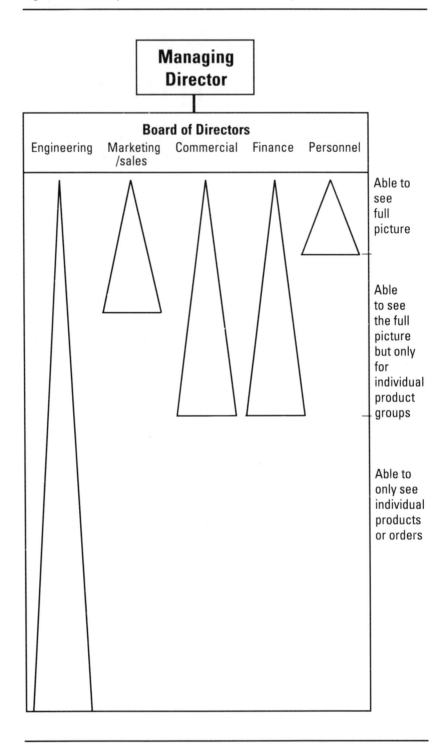

levels of authority and delegation, and what pieces of paper trigger activity. How does a new contract evolve into a series of purchase orders on subcontractors and suppliers? It is the kind of question that helps enlarge a sense of overall commercial awareness, it is said that a business breathes through its loop-holes. This is very true. It is vital to grasp how the whole of the operation is functioning.

Profit

Profit is the primary aim of any commercial business. Sound short-, medium- and long-term profit performance encourages new investment by shareholders and rewards existing share-holders with dividends and bonuses. Profit is the difference be-tween the cost of doing the job and the price for which it is sold to the customer. If costs can be reduced, profit is automatically increased and the converse equally applies. Although situations arise where a reduction in profit can lead to other benefits, the trade-off must be examined closely and carefully before a decision is made. In chess the queen may be sacrificed only if it is for the ultimate good of the king. So too in business. Profit may be sacrificed if there is some greater good to be accrued.

Cost

There are essentially two types of cost incurred by the business. Indirect (overhead) and direct costs. Indirect costs are recov-ered through the price of all products sold, direct costs are recovered through the price of the individual products to which they relate. The division of costs between overheads and direct varies from company to company. Find out what your company's accounting system is. Ask your commercial, finance or treasury department to explain it to you. Find out how your labour rates or job costs are calculated. From this you can determine which costs can be influenced by you with varying degrees of ease. Control of costs is the control of profit.

⑤ Payment and cash flow

Cash flow is vital to the business's well-being. Cash flow on the one hand is the outgoing of monies on salaries and wages, material and subcontract purchases and many other things; on the other hand is the incoming of payments from customers. Typically the two will not automatically balance, but nevertheless a balance must be achieved to ensure that a more-or-less neutral position is maintained. If you are involved in the preparation of proposals for stage-payments then find out how and when your costs will be incurred and disbursed and propose milestones that match these. Find milestones that will generate regular and frequent payments. Propose milestones that can be achieved as readily as possible and, as importantly, that can easily be proven to have been achieved. If you are involved in performing the work make sure that strategies, plans and practices for doing the work are geared towards meeting payment milestones. Make sure that control mechanisms are in place to permit changes in direction if progress in the work is steering away from achieving payment milestones. Look for opportunities to take milestones out of step if there is financial advantage. Find out if late or early achievement of milestones carries extra reward or penalty. Find out who is responsible for raising invoices and supporting paperwork and make sure that this is done promptly. Too many advantageous payment schemes have gone to waste for the sake of unclear responsibilities. Make sure that all the members of your team understand and pursue these objectives.

⑥ Estimating

Accurate estimating of time and costs is crucial to success. Inaccurate estimating can lead to losing orders or undertaking work on a loss-making basis. When an estimate is being prepared make sure that the task is accurately and comprehensively defined. If the task is one of design, for example, make sure that associated activities are clear. Documentation and data for development purposes is an obvious example, but what data? Has information for part numbering and cataloguing been thought of? If design-review meetings are necessary, how often, at what locations and with how many representatives? Without these bits of information an accurate estimate cannot be formulated. Make sure that there is no duplication or overlap. Application of project management and quality assurance at too many levels are classic examples of overestimating.

⑦ Products

The successful product is one that meets the customer's requirements, is cost-efficient to develop and manufacture, is flexible enought to permit and promote further development and sits in an expanding market. Ease of use, repair and maintainability, high reliability and availability, low cost and ready availability of spares are all obvious features of the ideal product. Nevertheless temptations exist to produce a technically 'superior' product that is gold plated or a more elegantly engineered solution. However, if these pursuits reduce the price competitiveness of the product or delay its entry into the market or cause late delivery then commercially they are a disaster.

⑧ The contract

It is important for all company employees engaged in the performance of the contract to understand the nature of the contract itself. The contract is the binding agreement between the buyer and the seller and should record, and as necessary explain, all the obligations and responsibilities of *both* parties. It is genuinely a two-way document. It not only sets out the customer's needs but also his obligations to make payments and to undertake various other actions. For example, the customer may be under obligation to take steps to protect commercially sensitive information supplied with the product. The contract will also set out rules and procedures to be followed in the event of disputes or questions of interpretation. Orders of precedence of documents may be given. For example, in the event of a conflict between the wording of a requirement specification and a test specification it is important to know which takes precedence.

As to the documents themselves, the contract will define all of the documents, whether incorporated physically or by reference, which together constitute the entire agreement between the buyer and the seller. For technical documents which can vary and progress in issue status the contract provides the baseline, either requiring contract amendment action or the adherence to a prescribed procedure before the new issues become binding. For these practical reasons it is highly desirable that those engaged in contract work become familiar with the contract and insist on explanation where the meaning of particular aspects is obscure. The contract 'terms and conditions' should not be dismissed as irrelevant.

⑨ Authority to commit

Once commitments are made they can be binding in contract and enforceable in a court of law. Find out who in the company has authority to make commitments, find out what limitations there may be and ensure that people outside the company with whom you deal understand the extent of your own authority and what you must do to acquire the company's approval in particular matters. Levels of authority and business functions possessing authority will vary according to the nature of the authority. For example, there may be difference in authority relating to offering of prices, contract negotiations, technical negotiations, amendments to contract, purchaser and so on. Binding undertakings can be made orally, directly or by phone, or writing. Be careful that comments, opinions, suggestions or recommendations could not be construed as contractual commitments. Where various meetings and committees occur in connection with a contract, ensure that the authority and terms of reference are clear and in line with the contract. Wherever and whenever correspondence is sent in connection with the contract make sure that clear reference is made to the contract and that references to meetings or discussions specify at least the date of the meeting or discussion. It is too often the case that regular correpondents dispense with these formalities, but in the event of difficulty, or dispute, the party which can demonstrate certainty will prevail. Where appropriate a written note should be kept of oral discussions. Work in concert with your commercial people and before any significant meeting discuss with them tactics and the extent of what can be discussed and agreed. If your customer or client tells you to proceed or says that contract or order cover exists or is about to be sent, check with your commercial people before starting work. If a customer can get his supplier to start work before paper contract cover exists he has the whip-hand.

(10) Definition of the work

Search for clarity in the contract definition of the work. Do not hesitate to challenge its adequacy, completeness or interpretation. Do not ignore what it says in the belief that you understand the customer's need despite the contract definition. This is a recipe for commercial disaster. The dangers are that the customer will not pay for extra or different work or that in the end the customer will say that he did want what was in the contract and not your interpretation. In your unique position you may be able to influence the customer to want more, but it must be on the basis that he will contract for more and pay more. The definition of what is required must extend to the description of the product, the required time-frame, the destination and any packaging and transportation requirements.

(11) Delivery

It is vital to understand from the contract the point at which delivery occurs and therefore the point at which risk of loss or damage passes from seller to buyer. This point can be at the supplier's works or at the customer's chosen destination or at any point in between. In the former case 'ex works' delivery means that the customer is responsible for collecting the goods from the supplier, including the provision of equipment and manpower, for loading on to delivery vehicles, and for transportation and insurance of the goods in transit. Too frequently an eager supplier will arrange to deliver goods despite the ex works nature of the contract, only then to find that he is responsible for loss or damage to the goods in transit. The correct delivery documentation is a crucial factor in achieving delivery. The contract should make this process clear and the company should ensure that a person is responsible for raising the paperwork. The project manager must ensure that goods offered for delivery and the associated paperwork are in line with the contract. Customers are entitled to reject paperwork. Deliveries rejected on these grounds can still be classed as delivery not made, possibly incurring penalties or other adverse consequences.

⑫ Customer discussions

Frequently in discussing possible requirements with customers attempts are rightly made to discover the amount of money the customer has available for the purchase, and other key facts. These efforts can be entirely wasted and opportunities lost if the status of the information is obscure or if the customer's procurement process is not adequately understood. For example, the following questions should be explored:

1) If the customer asks for a figure, find out whether he wants an estimate only for budgetary purposes or a quotation capable of formal acceptance and contract action.

2) Is his budget approved?

3) Is the particular procurement approved and if not what is the process and time-scale?

4) What influence does he have over the selection of a supplier? Who else has a say?

5) Does his budget include or exclude VAT?

6) When can contract action take place?

7) What validity is required for a formal quotation?

8) What commercial terms are standard or otherwise known that can affect price, e.g. warranty, delivery rate or time-scale, availability of interim payments?

9) What is the quotation evaluation and adjudication process?

10) Are existing or other suppliers preferred?

11) Does any bias exist in respect of subcontractors?

12) Which other companies are in the running?

13) Does the procurement form part of a larger system and/or are further procurements/developments planned?

It should not be forgotten that the customer is also talking to other potential suppliers and care should be taken not to such divulge information that could be useful to a competitor.

Whilst technical, engineering and project-management liaison with the customer is highly desirable, those involved should at all times take care in what is said and to discuss with the commercial department what may be **disclosed**.

Care must also be exercised in dealing with customer's requests for data or documentation relating to the contract. It should not be assumed that the customer has the right to any and all such material simply because he has the contract. In one particular case the customer believed he was entitled to any paperwork generated in the course of the execution of his contract. He was disappointed to find out otherwise. The customer's rights are limited to those defined in the contract. If the contract is silent on the matter then he has no rights.

(13) Intellectual property rights

Whether dealing with customers, suppliers, subcontractors or teaming partners, sight should not be lost of the key commercial significance of intellectual property and the rights at law and under contract that go with it. Intellectual property is ideas, inventions, technical know-how, associated drawings and other documentation, designs, trademarks, reputation and goodwill. Ownership or rights in these aspects which have been created largely at the company's expense must be bought or licensed by other parties. The fact that a customer has purchased some goods or even paid towards the cost of development does not necessarily mean that he acquires any rights in the intellectual property. The presumption must be that he acquires no rights at all unless they are specifically set out in the contract, and even then the exercise of those rights may be subject to special conditions (for example, no further disclosure of information) also set out in the contract.

Where companies wish to work together and exchange technical information and know-how for the pursuit of some joint enterprise the work will be preceded by the conclusion of a formal agreement between the companies setting out procedures and safeguards for the interchange and protection of information. It is important to be familiar with the contents of agreements so that the rules of engagement between the companies are clearly understood.

(14) Warranty

Where the customer is not happy or content with what he has bought he will want something done about it. It is a false assumption to believe that the supplier is automatically required to resolve deficiencies.

For example, these days everbody has heard of the requirements under the Sale of Goods Act for goods to be reasonably fit for purpose and of merchantable quality. However, this legislation was primarily introduced for the protection of the consumer who frequently has nothing more to go on than the appearance of the goods on a shelf. He has no opportunity to negotiate with the manufacturer regarding specification, performance, engineering standards, manufacturing standards or quality assurance. He buys on trust and the various consumer legislation is to provide rights in law if that trust is proven to be ill-founded. Thus in a contract where the parties do have the opportunity to negotiate standards, inspection rights, etc., the protection to the buyer under the Sale of Goods Act may very well not apply. Where the contract includes specific warranty provisions the rights of the buyer to require the seller to rectify defects free of charge may be limited in time and in scope. It is also important to recognise that, where the buyer has bought goods designed to his specific purpose, then once he has taken delivery he may find that the goods do not do what he had envisaged. The test here is whether or not the goods met the specification in the contract, and not the degree to which the customer is satisfied. The object is not to discourage attempts to cure customer dissatisfaction but to point out that what must be done at no cost to the buyer must be determined from the wording of the contract. The message to both buyer and seller is to be clear as to the obligations on both sides as set out in the contract.

One other aspect that is of practical interest in relation to warranty is the position of the main contractor in buying subcontract goods for onward shipment, whether buil into some into some larger article or system or not, to the end customer. Thought should be given to the period of time that will elapse between the subcontractor's delivery to the main contractor and the point at which the subcontract goods first come into use. First use may be by the main contractor in building and testing the larger article or first use may be by the end customer. In either case the period, particularly in a major contract, may be quite extended and could be greater period of time than that for which the subcontractor warrants the goods. In these circumstances the main contractor may consider dispensing with the subcontract warranty for a price reduction, or if the risk of the goods failing is considered significant then he may wish to negotiate with the subcontractor for the subcontract warranty to commence on first use rather than delivery. In this situation first use would need to be defined and probably an ultimate time-period for warranty-start agreed.

It is important for this type of consideration to be thought out when subcontractor/supplier requests-for-quotation are being prepared – an activity that many engineers become involved in.

⑮ Customer-furnished facilities, goods or information

Most major contracts involve at some stage or another a need for the customer to provide goods or information to the supplier or to make facilities or services available to him. The work under the contract may include the embodiment of customer-supplied items, or if the goods are to operate in conjunction with other equipment or systems of the buyer then the buyer must provide data (size, weight, shape, location, electrical and other inter-faces) regarding the other equipment or systems. If the work involves installation, on site testing or commissioning the buyer must provide ready access to the supplier not only for those particular activities but also for any necessary surveys or assess-ments in the earlier part of the contract. It is the responsibility make an impact on overall cost? Transport between his site and the buyer's or end-customer destination has to be paid for somehow. Also, the greater the distance the higher the chance of the engineer to make sure that all such needs are fully identified and defined to the maximum possible extent, including the dates by which the goods, information or services are needed. It is no good to rely on everybody assuming what is needed. It must be well defined and established in the contract as an obligation on the buyer. It is as important, if not more important, for the supplier to have these 'in-feeds' defined as it is for the supplier to have the outputs defined.

16 Selection of subcontractors (Fig. 11.2)

Identification of potential subcontractors and companies with whom association on a teaming basis would be beneficial is an activity rightly pursued by engineers in different functions. However, it is important to investigate the value of other companies commercially as well as technically. Having the right products is only a part of the story. The following lines of enquiry should be followed in conjuction with the commercial department:

a) Can the other firm comply with all the relevant aspects of the main contract, e.g. can quality-assurance, inspection, manufacturing and engineering standards be met? If not, there is a risk of being in breach of the main contract.

b) Does the firm have the project-management skills to enable it to support you and react effectively and efficiently to changes?

c) Are prices competitive; are they subject to premiums to cover matters mentioned in (*a*) and (*b*)?

d) Is the firm financially sound and large enough to take on the size and value of the job envisaged? If the firm is not financially stable or of sufficiently great financial resource there is a risk that your order will so strain him that he collapses, leaving you either late in contract performance or unable to complete at all.

e) Does the firm have sufficient capacity to handle the size of order contemplated? An overstretched supplier may not be able to keep to an agreed delivery programme or quality regime despite early optimism and reassurances.

f) Does the geographical location of the firm significantly

Figure 11.2: Subcontractor/supplier considerations

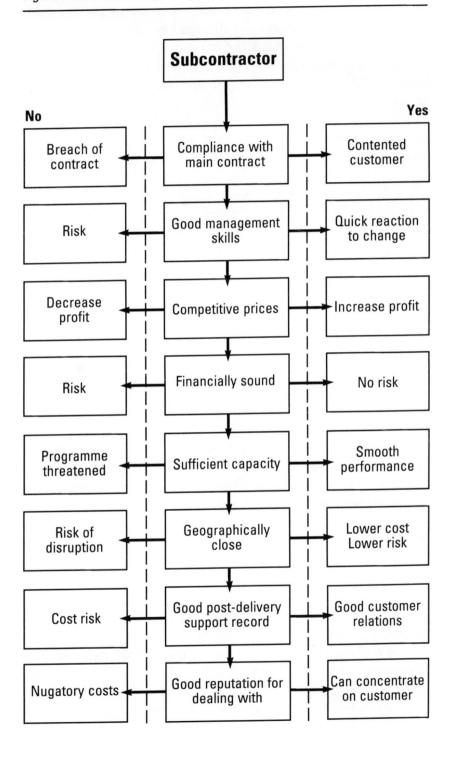

make an impact on overall cost? Transport between his site and the buyer's or end-customer destination has to be paid for somehow. Also, the greater the distance the higher the chance of disruption by weather or other factors such as rail strikes.

g) Does the firm have a good record of customer support in terms of spares and maintenance, etc?

h) Does the firm have a reputation for being easy to deal with? If the firm is difficult to deal with this will soak up valuable management effort that would be better deployed in looking after the customer's needs and the needs of the company.

As far as working with other companies on teaming arrangements is concerned, in addition to the foregoing points it is necessary to establish whether time and other factors make it feasible to set up the necessary arrangements.

17 Commercially sensitive information

The phrase 'commercially sensitive information' is often used, but what information is sensitive and from whom should it be protected? The simple rule is that all information with the exception of that printed in publications for general release is potentially sensitive. Particular areas of sensitivity are:

1) 'Technical details', particularly those that document or demonstrate ideas, inventions or know-how. Examples might include circuit diagrams, software source material, printed-circuit-board artwork.

2) Manufacturing information, again particularly those that document or demonstrate ideas, inventions or know-how. Examples would include manufacturing methods, process information and industrial engineering matters.

3) Production information: data relating to production plans, rates, targets, deliveries.

4) Product information: the progress of product development, plans for new or derivative products, dates for product launch, product reliability and performance.

5) Market information: the market types and regions, marketing tactics, level of marketing, details of which tenders will be bid for.

6) Sales information: sales made, sales targeted, order-

book information.

7) Financial information: costs or price breakdown, profit rates, labour rates, expenditure budget, cash flow, payment terms.

8) 'Legal' information: teaming or partnership relationships, contracts held, contracts details, customer complaints.

9) Customer/supplier information: customer/supplier lists, payment performance, credit-worthiness.

10) Company information: redundancies/recruitment planned, takeovers/mergers, closures, rationalisations.

Note that none of the headings above is 'commercial'. *All* of the areas are commercially sensitive and care should be taken not to divulge such information. The dangers include:

1) Straightforward commercial disadvantage; for example, a competitor learns of your costing/pricing system and wins business away from you.

2) Second-order disadvantage; for example, the mention of an idea allowing a competitor to patent the invention before you or the early disclosure of information which thus passes 'into the public domain' and prevents legal protection being established.

3) Misrepresentation; for example, if a customer is persuaded to buy goods on the strength of misleading information.

4) Breach of contract: disclosure of information (for example, details of contract held) could be breach of contract with the other party.

5) Breach of condition of employment: disclosure of information may be breach of your own contract of employment with your company.

The aim here is not to preach total silence but to point out the areas of risk and the potential pitfalls. Indeed the message is to seek out this information from your competitors but not to divulge your own. The boundary is admittedly somewhat grey. On the one hand it is correct to say that costing information should never be divulged, but on the other hand release of technical information to further encourage a potential customer whose appetite has been whetted by some initial material is usually a good thing. The question to be asked is whether or not the release of information should be preceded by a written agreement protecting the information once it has been divulged.

18 Negotiation

An entire chapter has been devoted to this particular activity. In anything other than routine business it is a fundamental truth that no difficulty of any significance was ever resolved by the two parties writing to each other. At the end of the day it is necessary to get out and meet across the table to fix the deal. This is the most sensitive of times and the engineer has a key role to play in its success.

19 Summary

The commercial engineer is the person who understands not only his particular expert subject but also the objectives and operation of his company such that, through efficient and effective performance of contracts at minimum risk and maximum profit, the business can be sustained and expanded to the benefit of shareholders, employees, the public and the nation at large.

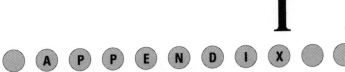

Case studies

Case studies are a most useful means to illustrate many of the principles and practical matters discussed in earlier chapters.

❶ Case study

'Blue Horizon' is designed to show the real difficulties into which both buyer and seller can get if there is uncertainty and lack of clarity and consistency between contracting parties. Indeed, not only between the contracting parties but also within each's own organisation, especially when it is not clear which department has what authority on different matters.

❷ Case study

'Engineering Consultancy Services' is an example of a real contract prepared by the buyer where the seller must decide whether the contract is appropriate, fair and acceptable on a clause-by-clause basis, and if not, to what extent he should seek to negotiate it with his customer, bearing in mind the opposed aims of minimising his own risk and keeping his customer sweet.

❸ Case study

'Widgets' is designed to show the muddle that both sides can get into if the contract is not precise in terms of delivery, acceptance, transfer of property and risk and warranty.

I

● ● ● ● ● C A S E ● S T U D Y ● ●

Blue Horizon

1. **The participants**

2. **The correspondence**

3. **Summary of events**

4. **Questions**

5. **Analysis**

① The participants

Macho Enterprises Plc: Prime contractor

Middleman Products Ltd: Subcontractor

Tiddler Components Ltd: Supplier

	Macho	*Middleman*	*Tiddler*
Contracts manager	Colin Macho	Chris Middleman	Clive Tiddler
Projects manager	Penelope Macho	Paul Middleman	Peter Tiddler
Purchasing manager	Brian Macho	Brenda Middleman	Bill Tiddler
Sales manager	Simon Macho	Steve Middleman	Sarah Tiddler

2 The correspondence

MIDDLEMAN
PRODUCTS LTD

Tiddler Components Ltd

For the attention of Sarah Tiddler   28 August

Dear Sarah

Blue Horizon – New Technology Widgets

You may know that we are expecting to receive an order from Macho Enterprises for the supply of NTW MK VII for their Blue Horizon project.

We have indicated that we would plan to include TCL LSI/3000/6F units in the NTW MK VII.

Please confirm that TCL could supply the items and if so let me know what the cost would be for 5000 off. Please reply by 3 September.

Yours sincerely

Paul Middleman
Project Manager

Tiddler Components Ltd

Telex

Middleman Products Ltd

For the attention of Paul Middleman - Project Manager

NEW TECHNOLOGY WIDGETS

Thank you for your letter.

Cost for 5000 off LSI/3000/6 units is £100 each. Delivery 4 weeks f.r.o. subject to standard conditions.

Are we still ok for golf Saturday week?

Regards

Sarah Tiddler
Sales Manager

2 September

MIDDLEMAN
PRODUCTS LTD

TELEX

Tiddler Components Ltd

For the attention of Clive Tiddler – Contracts Manager

BLUE HORIZON

Macho have secured the order and given us an intention to proceed.

Delivery of TCL LSI units must start 1 November if we are to meet Macho programme.

Purchase order in preparation.

Regards

Paul Middleman
Project Manager

2 October

Tiddler Components Ltd

Telex

Middleman Products Ltd

For the attention of Paul Middleman - Project Manager

BLUE HORIZON

Congratulations and thank you for your telex 2 October.

Work is in hand and we look forward to receipt of formal purchase order.

Regards

Clive Tiddler
Contracts Manager

3 October

Tiddler Components Ltd

Quotation

Middleman Products Ltd

For the attention of Brenda Middleman

Dear Brenda

BLUE HORIZON

Further to Paul Middleman's 2 October ITP we are proposing to deliver 5000 off LSI/3000/6 units on 15th November.

Price each is £120 plus VAT invoiceable following delivery.

Yours sincerely

Clive Tiddler
Contracts Manager

3 October

Tiddler Components Ltd

3 October

Middleman Products Ltd

For the attention of Paul Middleman – Project Manager

Dear Paul

BLUE HORIZON

I am glad we're straight on the Blue Horizon job. The 5000 off should be with you on 3 November providing you can let me have the electrical interface data by 5 October.

Yours sincerely

Peter Tiddler
Project Manager

MIDDLEMAN PRODUCTS LTD

PURCHASE ORDER

TO: Tiddler Components Ltd   Order No A123/4

Please supply:

QTY	SPEC	DESCRIPTION	PRICE
5000	LSI/3000/6E	LSI FOR NTW MK VII	£100 each

Delivery: 500/week commencing 1 November

Packaging: Retail trade

Terms: Standard

Acceptance: Please confirm acceptance

Signed ....

Brenda Middleman
Purchasing Manager

6 October

Tiddler Components Ltd

10 October

Middleman Products Ltd

For the attention of Brenda Middleman – Purchasing Manager

Dear Brenda

ORDER NO A123/4

Thank you for your purchase order dated 6 October which is acceptable although the price and delivery require revision in line with our 3 October quote.

Yours sincerely

Clive Tiddler
Contracts Manager

**Macho
Enterprises Plc**

20 November

Middleman Products Ltd

For the attention of Steve Middleman – Sales Manager

Dear Steve

BLUE HORIZON

Unfortunately Blue Horizon has been cancelled and we cannot go ahead with the New Technology Widget order.

Yours sincerely

Penelope Macho
Project Manager

③ **Summary of Events**

Date	From	To	
28 Aug.	Middleman Project	Tiddler Sales	Request cost of 5000 off LSI/3000/6F
2 Sept.	Tiddler Sales	Middleman Project	Cost £100 each 5000 off LSI/3000/6 Delivery 4 weeks f.r.o. standard conditions
2 Oct.	Middleman Project	Tiddler Contracts	Advises Middleman have intention to proceed, says LSI delivery must start 1 Nov.
3 Oct.	Tiddler Contracts	Middleman Project	Work is in hand. Looking forward to formal purchase order.
3 Oct.	Tiddler Contracts	Middleman Purchasing	Quotes £120 each for LSI/3000/6. 5000 off to be delivered on 15 Nov.
3 Oct.	Tiddler Project	Middleman Project	Promises 5000 off on 3 Nov. if electrical interface data provided by 5 Oct.
6 Oct.	Middleman Purchasing	Tiddler	Purchase order 5000 off LSI/3000/6E. £100 each 500/week commencing 1 Nov.
10 Oct.	Tiddler Contracts	Middleman Purchasing	Accepts purchase order subject to price and delivery revisions.

4 Questions

1. Does a contract exist between Middleman and Tiddler?

2. When did the contract come into being?

3. What is to be supplied and when?

4. What status does Peter Tiddler's letter about electrical interface data have?

5. What is the price?

6. What are the terms and conditions of the contract?

7. Does cancellation of Blue Horizon give Middleman an opportunity to cancel the order with Tiddler?

⑤ Analysis

Does a contract exist between Middleman and Tiddler?

The primary criteria for the existence of a contract are:
1) Legal capacity of the parties.
2) The contract must be legal and possible.
3) Intention to create legal relations.
4) Consideration.
5) Offer and acceptance.

In this instance it is safe to assume that Middleman and Tiddler have the legal capacity to enter into contracts. Their limited-company status indicates incorporation within the Companies Act and clearly they intend to carry on the business of trading. Criterion no. 1 is satisfied.

The contract – if it is such – is legal and possible. The sale of the LSI units is not illegal either in the criminal sense or in the civil sense. That is to say it can be assumed that the goods in themselves are not illegal (as prohibited substances may be), that the act of selling them is not illegal (as sale of firearms may be) and that Tiddler actually has the right to sell them, i.e. that they do not belong to somebody else. The contract is also possible in the sense that a 'contract' for a perpetual-motion machine, for example, would not be possible. Thus the contract is legal and possible and criterion no. 2 is satisfied.

The question of an intention to create legal relations is slightly more difficult. In principle there is no doubt that Middleman intended to buy LSI units from Tiddler to satisfy the Macho order. Clearly Tiddler were keen to sell. However, the early correspondence could not be construed as conveying a specific intention at that time. For the time being it is assumed that the intention did exist on both sides although the timing of events, as will be seen later, is important. Criterion no. 3 is satisfied.

Consideration: in this case the obligation to pay money for the goods is clear. Each side may have a different view of the price but it is certain in principle that Middleman intended to pay. Which is the correct price is discussed later. If the lower price stands and if that would mean Tiddler selling at a loss the consideration is still sufficient. Criterion no. 4 is satisfied.

Offer and acceptance is far from clear. At no time is there a clear offer met with an unqualified acceptance. On top of that it is not entirely certain what the subject of the contract is. LSI/ 3000/6, LSI/3000/6E, LSI/3000/6F are all mentioned. As has been said, there is no agreed price and delivery is also uncertain. In the extreme a court may decide that the contract is void in the face of all this uncertainty. In practice it is most likely that the parties will resolve the uncertainties, supply of the goods will go ahead and payment will be made. In these circumstances there is said to be a 'contract by performance' and the offer acceptance is deemed to have occurred. Criterion no. 5 is satisfied.

When did the contract come into being?

In the final paragraph above it was concluded that, if the work had proceeded, then a contract did exist and the circumstances of 'contract by performance' meant the contract probably would come into being when delivery was made. It is important to use the word 'probably' as there are no right, or certainly no exact, answers to some of the questions arising from the particular sequence of events. Had the realisation that Blue Horizon was cancelled come before delivery had started then it was still possible that, taken together, the full set of correspondence would constitute a contract.

Most probably it could be said that Tiddler's 3 October letter saying work is in hand in reply to Middleman's 2 October letter stressing the importance of starting delivery on 1 November created a contract, especially as Middleman did nothing afterwards to disabuse Tiddler from the belief that they had been asked to start.

The later issue of the purchase order and subsequently its qualified acceptance are formalities only, unless it is clear – and provable – that Tiddler should have known that only a properly authorised purchase order could commit Middleman to contract.

What is to be supplied and when?

Assuming then that, whether or not a contract existed in the strict legal sense during the early part of the events, both parties proceeded in the belief that a contract existed, the question arises as to what should be supplied and when.

If in practice there was no significant difference between the LSI/3000/6, LSI/3000/6E and LSI/3000/6F it may not matter what was delivered. If E and F are versions of a 6-series family having no form, fit or function difference in so far as Middleman's New Technology Widget was concerned, there should be no cause for complaint from Middleman. If there are differences then both sides have a good argument. Tiddler were consistent between 2 September and 3 October in offering to supply the LSI/3000/6. Although the initial request for quote had mentioned the 6F, Middleman did not query the two Tiddler letters. However, Tiddler failed to query the number in their response to the purchase order which referred to the 6E. This misunderstanding is potentially the most serious of the defects in this chain of events. If Middleman genuinely needed the 6F as per their enquiry or the 6E as per the purchase order and Tiddler tried to deliver the 6 version then the delivery would be rejected – causing Tiddler severe problems – and Middleman could fail to perform their contract (assuming Blue Horizon had not been cancelled) with Macho.

The question of time of delivery is equally fraught. This time both parties were consistent. Middleman twice said that delivery must start on 1 November and Tiddler mentioned only 3 and 15 November (this in itself is very poor, two Tiddler representatives independently giving different dates). Tiddler also referred to this in the response to the purchase order and indeed their initial offer mentioned 4 weeks from receipt of order and at that time there was sufficient time for Middleman to place the order. Tiddler's later letter on 3 October is also consistent with this.

One problem is that if Tiddler agreed that the 2 October letter was the go-ahead, then the wording of that letter suggests that 'time was of the essence' to Middleman. In that case Tiddler would be in fundamental breach of contract if they did not deliver on the 1 November entitling Middleman to cancel the contract.

A further complication is the mention on the purchase order of delivery in batches. Although Tiddler queried the delivery it is not absolutely clear if they were querying the batch requirement or the start-date or both. For Tiddler to plan for single delivery but in practice to send batches could give them cash-flow and storage problems. For Middleman to plan batches but receive single delivery would give cash-flow and storage problems to Middleman.

The balance of weight appears to be on Tiddler's side and thus a single delivery on 3 November would have been correct, but what of the electrical interface data problem?

What status does Peter Tiddler's letter about electrical interface data have?

The letter is certainly consistent with the offer of delivery 4 weeks from receipt of order if it is assumed – as Tiddler did – that the 2 October letter from Middleman was the instruction to proceed. It clashes with the 3 October letter from Clive Tiddler which mentioned the 15th and did not refer to the interface data. Perhaps the contracts man was assuming that the data would not arrive by 5 October or perhaps he just did not want to commit his company to such a tight delivery obligation given the risk, mentioned above, that the contract could be interpreted as 'time is of the essence'. In any event the dependency on receiving the interface data was not mentioned by Tiddler in either quotation or specifically in the response to the purchase order.

If delivery by Tiddler was delayed beyond 15 November and Middleman did not provide the electrical interface data on time it is unlikely that Tiddler could successfully argue that the delay was the fault of Middleman.

What is the price?

Assuming that there is no price difference between the LSI/3000/6, LSI/3000/6E and LSI/3000/6F then is the price £100 each or £120 each?

The initial enquiry from Paul Middleton asks for the cost. It is not clear whether he meant to ask for an estimate (for guidance purposes) or a quotation (capable of acceptance). In this context, asking for cost does not help. A price from Tiddler may be a cost to Middleman. Sarah Tiddler's reply is no more helpful, although when Clive Tiddler did quote formally on 3

October nobody from Middleman queried the discrepancy.

The difficulty may be resolved if Middleman knew that either Sarah Tiddler did not have authority to quote or that Tiddler quotations are only official if communicated on formal 'quotation' paperwork. Tiddler did comment on the price shown on the purchase order and so the £120 price probably stands, although perhaps Middleman might push Tiddler for a reduction if in good faith they had already used £100 in quoting to Macho.

What are the terms and conditions of the contract?

This is almost impossible to say. Tiddler's standard conditions were referred to in Sarah Tiddler's letter, but for the financial reasons discussed above Tiddler would prefer that letter not to be formal. However, Clive Tiddler's quotation – which Tiddler would prefer to be the formal quotation – did not mention terms. The purchase order referred to Middleman's standard terms. Tiddler did not comment on this in responding to the purchase order. The balance is slightly in Middleman's favour.

In practice, if neither set of conditions contained anything that gave rise to dispute the parties would probably ignore each other's conditions and thus it would never be known as to which applied. The majority of business transactions are carried on in this way and hence occasionally it is wondered why time is devoted to the terms and conditions. The answer can be seen in the parallel with insurance – it is there as a safeguard against the small chance of something going wrong, and when something does go wrong the safeguard must be sufficient and appropriate.

Does cancellation of Blue Horizon give Middleman an opportunity to cancel the order with Tiddler?

On the basis that a contract between Middleman and Tiddler exists, then the Blue Horizon cancellation notified by Macho has no effect on the Middleman/Tiddler relationship. If Middleman could establish that it is their standard conditions that apply then it is likely that those (i.e. the conditions favourable to the buyer) would include a right for Middleman to cancel the contract for their own convenience. If not, Tiddler would be entitled to insist that Middleman negotiate a settlement relating to the cost or value of work done, cancellation charges and lost profits if the contract is to be prematurely terminated.

Summary

The sketchy example used may appear to be exaggerated but such events do happen, particularly where matters are complicated – as they are in practice – by the involvement of more people, more correspondence, telephone and other discussions, interdepartmental communication and other problems, etc., etc. The main points illustrated are:

a) the need for certainty, clarity and consistency in all communication,

b) the need for clear co-ordination between different departments,

c) the need for clear understanding on all sides of who is authorised to do and say what.

II

C A S E   S T U D Y

Engineering Consultancy Services

1. **Introduction**

2. **The subcontract**

3. **Analysis**

4. **Summary**

① Introduction

The subcontract agreement is a proposed agreement between a supplier of engineering consultancy services and a company who needs the assistance to satisfy a contract with his client.

The schedules A, B, C and D referred to are not included as the commercial principles can be satisfactorily demonstrated without them.

To assist in understanding, the terminology in the Analysis is varied according to the context. Thus the company buying the services is referred to as the 'main contractor' or the 'buyer'. The consultancy firm is referred to as the 'subcontractor', 'seller' or 'supplier'. The subcontract agreement is referred to as the 'subcontract' or the 'contract'.

② The subcontract

1. The agreement

1.1. Background

1.1.1. This Agreement is supplemental to a contract as referenced in Schedule A (hereinafter referred to as 'the Main Contract') made the 28th day of Apil 1983 between Blinkers Ltd (hereinafter referred to as 'the Client') and the Contractor which provides for the provision of Engineering Consultancy on the terms and conditions therein contained.

1.1.2. The Subcontractor has agreed with the Contractor that it will carry out the services and provide the articles and facilities detailed in Schedule B hereto (hereinafter referred to as 'the Services' and 'the Articles' respectively) in the manner and at the time specified therein in connection with the project to be carried out under the Contract (the 'Project') on the terms and conditions herein referred to.

1.1.3. In this Agreement words and expressions shall except as otherwise provided have the same meanings as are respectively assigned to them in the Main Contract.

1.2. The services

1.2.1. The Contractor shall employ the Subcontractor to carry out the Services and provide the Articles in accordance with the criteria set out in Schedule B and subject to the terms and conditions of this Agreement and the conditions of the Main Contract so far as they are applicable. In the event of inconsistency or conflict the conditions in the Contract will at all times prevail, unless otherwise expressly agreed in writing. Failure by the Subcontractor to meet the criteria set out in Schedule B shall, for the purposes of Clause 3.14 be constructed as a serious breach.

1.2.2. In addition to supplying the Services and Articles set out hereto the Subcontractor shall carry out other work in connection with the Project as the Contractor shall specify, provided the price is agreed by the Contractor prior to the work commencing and the work to be performed by the Subcontractor will be carried out in accordance with the terms of this Agreement.

1.3. Project management

For the purposes of the smooth handling and implementation of the Main Contract and the co-ordination of the Project, the Contractor shall appoint a project manager (the 'Project Manager') to whom shall be delegated the day-to-day responsibilities of managing the Main Contract and who shall be responsible for the co-ordination of subcontractors. The Subcontractor shall also appoint a representative through whom contact can be made. The Subcontractor shall obey all reasonable and lawful instructions of the Project Manager as to the manner in which the Services and Articles are to be supplied.

1.4. The project plan

These are shown in Schedule D (attached) and it is agreed that both the Contractor and the Subcontractor will work to these arrangements. The parties to this Agreement will report progress to the other at agreed intervals.

Terms and conditions
2. Pricing

2.1. Price and payment terms

The consideration due (hereinafter referred to as 'the price') for carrying out the Services and/or supplying the Articles shall be as specified in Schedule C hereto and shall become due and payable by the Contractor to the Subcontractor in accordance with the provisions therein contained.

2.2. Delay in payment

Payment by the Contractor to the Subcontractor may be delayed as a result of the Subcontractor's failure:

a) to send on the day of despatch for each consignment, advice of despatch and invoice, or

b) to send a monthly Statement of Account quoting the invoice numbers applicable to each item thereon, or

c) to mark clearly the Contractor's Project Number/Main Contract Reference on any packages, packing notes, advice notes, invoices, monthly statements and all other correspondence relating thereof.

3. General

3.1. Assignment

a) The Subcontractor shall not assign the whole or any part of the benefit of this Agreement nor shall the Subcontractor sublet the whole or any part of the Services or Articles without the prior written consent of the Contractor.

b) 'Subletting' in this Clause shall include placing an order for the supply of articles and/or services in connection with the Services and/or the Articles.

3.2. Duration

This Agreement shall start on the Commencement Date and shall continue until *either* all the obligations of both parties have been fulfilled *or* the Agreement is terminated in accordance with clause 3.14 *or* by mutual written Agreement, whichever occurs first.

3.3. Liability

a) The Contractor shall indemnify the Subcontractor in respect of any liability for death of or personal injury to any person caused by the Contractor's negligence.

b) The Contractor shall not in any circumstances be liable to the Subcontractor whether in contract, tort or otherwise for any consequential or indirect loss or damage, howsoever arising and of whatsoever nature including (without limitation) loss of profit, loss of contracts, loss of operation time, loss of computer-held data, loss of use of any equipment or process or any other form of loss whatsoever (whether or not similar to some or any of the foregoing) suffered or incurred directly to indirectly by the Subcontractor.

c) The liability of the Contractor to the Subcontractor for direct loss or damage whether in contract, tort or otherwise arising out of or in connection with its performance or its total or partial failure to perform in accordance with the terms of this Agreement shall in respect of any one incident or series of incidents attributable to the same cause be limited to and shall not in any circumstances exceed the lesser of the sum of £300,000 (three hundred, thousand pounds sterling) or twice the Price under this Agreement.

3.4. Indemnity

a) The Subcontractor *hereby acknowledges* that any breach by it of this Agreement may result in the Contractor committing breaches of and becoming liable for damages under the Main Contract and other agreements (in like form to this Agreement and otherwise) made by the Contractor in connection with the Project and all such damages, losses and expenses are hereby agreed to be within the contemplation of the parties as being probable results of any such breach by the Subcontractor against all such breaches and damages as aforesaid.

b) Without prejudice to the indemnities above in favour of the Contractor, the Subcontractor shall take out all necessary public, product and professional liability insurance with an insurance company or companies of repute to cover the liabilities assumed by it hereunder and shall on request make available such policies for inspection by the Contractor or the Client as the case may be.

3.5. Intellectual Property Rights

The Subcontractor recognises that all Software, listing and associated documentation and manuals supplied, are the Contractor's confidential information and all Intellectual Property Rights therein are and shall remain vested in the Contractor. The Subcontractor shall not, without the Contractor's prior consent in writing, adapt, modify, or make any copies of or divulge to any third parties other than the Client the said confidential information and shall so bind its directors and employees, provided that the Subcontractor may for its own use make copies as may be necessary for the proper operation or security of its business.

3.6. Ownership of Results

All rights in the result of the work performed by the Subcontractor in the course of this Agreement shall forthwith be communicated to, and shall belong exclusively to, the Contractor. If requested by the Contractor, the Subcontractor will agree to do all things necessary at the Contractor's sole cost to obtain where possible letters patent, registered designs, copyright or like industrial property in relation to any process product, concept or writing developed or produced by the Subcontractor in the performance of such work.

All Articles prepared or developed by the Subcontractor under this Agreement including maps, drawings, models and samples, shall become the property of the Contractor when prepared, whether delivered to the Contractor or not, and shall, together with any Articles furnished to the Subcontractor by the Contractor, be delivered to the Contractor upon request and in any event upon termination of this Agreement. This provision will also apply to work completed after the date of termination of this Agreement, but restricted solely to work which forms part of the Project. Rights in concepts, products, processes or writings already developed by the Subcontractor prior to the date of this Agreement, remain the property of the Sub-contractor.

3.7. Drawings

All drawings, specifications and similar data issued in connection with this Agreement are to remain the Contractor's property and must be surrendered to the Contractor upon completion or termination. They must be used solely by the Sub-contractor in aid of the manufacture and supply of the Articles and for no other purpose whatsoever excepting with the Contractor's prior written consent.

3.8. Delivery

a) The Articles shall be delivered not later than, nor more than 28 working days prior to any dates specified in Schedule D and delivery shall not be deemed effected until the Articles have been received at the destination specified.

b) The Contractor reserves the right to terminate this Agreement or any part thereof and to claim damages in the event that the Articles or any part thereof are not received by the contractor or the Services are not completed by the date or dates agreed in Schedule D or the Articles or the Services are not in accordance with this Agreement or any part thereof.

3.9. Poaching staff

Both parties agree, that during the period of this Agreement and for 18 months after its conclusion, they will not directly employ any of the staff of the other party at any time engaged in the pursuance of this Agreement without prior written consent by that other party which shall not be unreasonably withheld.

3.10. Confidentiality

3.10.1. Each party undertakes to keep and treat as confidential and not disclose to any third party other than the Client any information relating to the business or trade secrets of the other nor make use of such information for any purpose whatsoever except for the purposes of this Agreement provided that the foregoing obligation shall not extend to information which is:

a) in or comes into the public domain other than by breach of this Agreement;

b) in the possession of one party prior to receipt from the other party;

c) received bona fide by one party from a third party not receiving the information directly or indirectly from the other party.

3.10.2. However, nothing in this Agreement shall operate so as to prevent either party or any of its staff from making use of know-how acquired, principles learned or experiences gained during the execution of the Agreement. This clause is binding on all parties during the Agreement and for a period of three (3) years after termination and each party shall so bind its directors and employees.

3.11. Force Majeure

Neither party shall be under any liability to the other party for any delay or failure to perform any obligation hereunder if the same is wholly or partly caused, whether directly or indirectly by, circumstances beyond its reasonable control.

3.12. Publicity

No publicity or advertising shall be released by either the Contractor or the Subcontractor in connection with this Agreement without the prior written approval of the other which shall not be unreasonably withheld.

3.13. Alteration of Agreement

No alteration, modification, or addition to this Agreement,

nor any waiver of any of the terms hereof shall be valid unless made in writing and signed by the duly authorised representatives from both parties.

3.14. Termination

If either party commits any serious breach of its obligations hereunder and fails within sixty (60) days of written notice to remedy the same, the other party may forthwith, by notice in writing, terminate the Agreement without prejudice to any other rights which may have accrued to it hereunder.

3.15. Notices

All notices shall be in writing and shall be directed to the Contractor or to the Subcontractor each at its respective address shown on the face of this Agreement or to such other address as the recipient may from time to time specify in writing.

3.16. The Agreement

a) The entire Agreement between the Subcontractor and the Contractor with respect to the subject matter herein is contained in this Agreement and Schedules hereto and supersedes all previous communications, representations and arrangements, either written or oral and the Contractor hereby acknowledges that no reliance is placed on any representation made but not embodied in the Agreement.

b) The Agreement shall be construed and governed in accordance with English Law.

Signed.........................   Signed.............................
For and on behalf of   For and on behalf of
....
Date............................   Date................................

③ Analysis

1. The agreement

1.1. Background

1.1.1. It is helpful to both parties to the contract for a little of the background to be included by way of introduction. Theoretically it is irrelevant and should not be included as the contract in itself should be sufficiently clear and precise as to the obligations of and benefits to each side such that a question of interpretation can be settled by an independent third party – an arbitrator or court perhaps – regardless of the context. However, the buyer in including reference to his customer is beginning to build a connection between his supplier and customer.

1.1.2. Here again a long, rambling legalistic sounding sentence is saying that the supplier will perform the subcontract in line with the requirements of the main contract. Theoretically unnecessary, as the subcontract should be self-sufficient but the buyer will include it in case there is some aspect of the main contract which he has unwittingly failed to cover in the subcontract. The seller should avoid all this if he can.

1.1.3. This provides further emphasis that the subcontractor is hooked into the main contract. It may be all very well in principle that end-customer, main contractor and subcontractor share common expression, and meanings but in practice how can the subcontractor know what meanings are understood between main contractor and end-customer? The expression 'except as otherwise provided' means that this general rule (of meanings from the main contract) can be varied by specific statement(s) within the subcontract.

1.2. The services

1.2.1. On the face of it this is a reasonable enough clause, providing as it does that the two parties must agree a price for additional work before that work is put in hand – and if so that it will meet the terms of this agreement, i.e. that it will meet the needs of the main contract. Read carefully, it can be seen that the clause effectively puts an obligation on the subcontractor to carry out additional work if the main contractor requires it. In practice this may not suit the subcontractor in terms of allocation of resources, etc., when he has no idea what extra might be required and when. The subcontractor could always delay discharging the obligation by arguing over the price.

1.3. Project management

This clause is fair enough. It establishes that points of contact and communication channels will be identified to ensure smooth working. In practice all will depend upon the reasonableness of the two project managers and their staff. Taken to the extreme the main contractor project manager could make discharge of the subcontract difficult, although the subcontractor would rely on the phrase 'reasonable instruction'.

1.4. The project plan

The wording here causes no difficulty. The subcontractor must make sure that the project plan is sufficiently clear as regards his activities as, once again, the main contractor is incorporating main-contract obligations into the subcontract. Both sides will seek to ensure that the rate and content of progress-reporting provides adequate communication with intolerable-cost limitations.

2. Pricing

2.1. Price and payment terms

In principle this clause is perfectly acceptable. By reference to Schedule C, the main body of the subcontract is identifying and *linking* price and payment. Too often, people are concerned with getting the price agreed without giving thought to the timing and mechanism of payment. One interesting point is that the clause makes it clear that the consideration comprises solely of the price. That is, if there are other obligations upon the buyer of value to the seller these would not be deemed to form part of the consideration.

2.2. Delay in payment

From the main contractor's point of view all of this is entirely necessary. The clause identifies those things which in practice will identify and verify to the main contractor's payment systems that payment claimed under the subcontract is valid, correct and due. If his system cannot make this identification and verification he cannot pay, although the subcontractor may indeed have fulfilled the primary obligation of providing the goods or services.

What the clause neglects to say is how quickly payment will be made once the subcontractor has remedied the defect in these minor respects. Indeed, the clause also fails to oblige the main contractor to promptly notify the subcontractor that payment has been delayed for one or more of these reasons.

3. General

3.1. Assignment

The effect of this clause is to say that the main contractor has let the subcontract to the particular subcontractor and that he wants no other party involved without first giving his written agreement. That is, the subcontractor cannot further subcontract the work or promise the benefits (the money) or the subcontract to anybody else.

3.2. Duration

This is an even-handed statement saying, in conjunction with 3.14, that neither side can get out of his obligations unilaterally. This is slightly unusual as the buyer frequently reserves the right to bring the arrangement to a premature end for his own convenience.

3.3. Liability

a) Ostensibly this is a generous provision for the main contractor to include. In practice it is only a statement of the fact of law. The main contractor is liable if his negligence causes death or personal injury and that liability could not be eliminated by a statement of exclusion in the contract.

b) Here the main contractor is saying that he will not be liable to the subcontractor for the long list of possible *indirect* losses mentioned. Legally this is acceptable although the subcontractor should seriously examine all the categories and decide which are both likely and serious and attempt to negotiate their removal from the list of exclusions.

c) This clause relates to the main contractor causing direct loss to the subcontractor; for example, a main-contractor vehicle damaging subcontractor property in collecting goods. The financial limitation is likely to be driven by the insurance cover held by the main contractor. The lesser of the sum of £*xx* or *Y* times the price of the subcontract is a common arrangement and is acceptable. Of course where the price of the subcontract is not known at the outset, it is a question of judgement and negotiation as to the size of the *Y* factor.

As a whole, this clause sets out to protect the main contractor by elminating or limiting (where these are permitted by the law) his liability to the subcontractor for his negligence or other deficiencies. The subcontractor may seek to agree that the clause is revised to give it mutuality in effect.·

3.4. Indemnity

a) Following on from 1.2.1 this clause really hammers it home that the subcontractor is only there to help the main contractor perform its contract and that the subcontractor is fully and solely responsible for main contract breaches caused by the subcontractor.

On the principle of 'for want of a nail the battle was lost' the main contractor is right to be very concerned that, having taken the decision and the risk to subcontract, he will not be seriously let down. On the other hand, the subcontractor can argue that he cannot possibly know sufficient of the main contract, that as he is only the subcontractor (i.e. not enjoying the benefits of the main contractor) and that as he is working at the instruction of the main contractor's project manager he should not take on that liability.

b) This is a natural and reasonable precaution for the main contractor to secure. If his subcontractor fails in certain ways it is one thing to have the right to sue him, but if this does not yield recompense – because the subcontractor simply goes bankrupt – the right is of no commercial value. Thus the main contractor will want to know that the subcontractor is protected by adequate insurance policies. If the concern is great enough the main contractor may look to the subcontractor to have him identified in the insurance policy itself as having an interest.

3.5. Intellectual property rights

The key words here are 'supplied' in the second line and

'are and shall remain' in the fourth line. Together it can be interpreted that the clause is addressing rights in documentation issued by the main contractor to the subcontractor for the purpose of the work under the subcontract. As the clause invites interpretation it is wise for the subcontractor to seek confirmation of this interpretation. Assuming this is correct the subcontractor may seek an indemnity from the main contractor against legal action by third parties in respect of IPR infringement. After all, the subcontractor has only the main contractor's word that the IPR belongs to him. If the main contractor is telling the truth there should be no problem in him giving that indemnity. If he will not readily give the indemnity then the subcontractor would be right to have some concern.

3.6. Ownership of results

Provided that the subcontractor is content as a matter of policy to have no rights in design work performed under contract then this clause is acceptable to him. In practice the difficulty is in the merging of ideas and information and thus the separability of intellectual property belonging exclusively to each of the two parties is impossible to achieve.

3.7. Drawings

The important point to make is that this clause is addressing ownership physically of the drawings and other material rather than the intellectual property enshrined within them which is covered by 3.5.

3.8. Delivery

a) This relatively brief clause is to some degree unclear in so far as it is not certain whether the main contractor or the subcontractor is actually resposible for delivery to destination. It only goes so far as saying that for contractual purposes delivery has not occurred until goods have been received at their destination. Transportation and insurance in transit are not mentioned.

b) Unless the contract says so specifically (as it does here) the buyer may not be entitled to terminate if the supplier fails to deliver *exactly* on time or *exactly* to specification. That is, in the normal course of events, late supply and/or delivery of goods not exactly to specification would entitle the buyer only to sue for the resultant damages (if any) for non-conforming performance in terms of time or specification. It is for the supplier to judge how likely these possible events are and to react accordingly. For example, if he believes that late delivery is a distinct possibility and that resulting from that the buyer is likely to suffer real damage (i.e. of genuine financial impact) then he may wish to negotiate a liquidated-damages arrangement. The advantage to the supplier is that, worded correctly, such an arrangement would introduce an absolute maximum liability for late or non-delivery. On the other hand, he could find himself making liquidated-damages payments to the buyer in circumstances where, in practice and although the supplier was late, the buyer would not actually pursue the defaulting supplier for redress.

3.9. Poaching staff

This type of clause tends to be seen where the type of work involves a fairly close technical relationship between the parties whereunder each may identify in the other's employees potential recruits. It should be noted that as far as such a clause can go is to restrict the activities of the parties in so far as recruitment processes are concerned. It is probably illegal – or at least unenforceable – if the clause were to seek to prevent the parties from actually employing members of each other's staff.

3.10. Confidentiality

3.10.1. These days, such statements of restriction are quite common. In practice they place upon the other party's employee's obligations equivalent to those upon employees of the company under each employee's contract of employment. That is, each employee must keep confidential the secrets of his employer and the secrets of those firms with whom his employer contracts.

Subparagraph (*a*) is there to point out that information already available to the public at large cannot suddenly be made confidential simply by the wording of a contract between two parties.

Subparagraph (*b*), similarly, points out that if firm *X* already holds certain information, that information cannot become subject to the conditions of confidentiality simply because it is re-transferred to him from *Y*.

Subparagraph (*c*) covers the final permutation of events whereby one party might independently and legitimately receive information from a third party which, had it not been so received, the other party might have had some claim that it should be held in confidence.

3.10.2. The first sentence is quite reasonable and equitable although, in permitting the subcontractor freedom to use knowledge gained, it conflicts with clause 3.6. This is something which the subcontractor should query. Clearly it is in his best interest to have the former statement prevail.

The second sentence is probably meant to apply to the whole of clause 3.10 rather than just the first sentence of 3.10.2. It is important for the subcontractor to establish exactly and unequivocably what his rights and obligations are.

3.11. Force Majeure

Always the two questions that each party should ask itself in considering going for the inclusion of a *force majeure* clause are (i) how likely delay/default is and its potential consequences on the buyer and (ii) the degree to which it is desirable at the outset to argue out what precisely are the circumstances which, for the purposes of the clause, provide excusable reasons for delay/default.

From the seller's point of view he must also decide whether to propose liquidated damages.

To agree at the outset what the buyer's maximum damage could be is to limit the seller's exposure. However, in practice he could find himself making liquidated-damages payments to the buyer in circumstances where, but for the inclusion of liquidated damages, the buyer would not actually have pressed the Seller for recompense for the delay.

At least this particular *force majeure* clause is mutual in its effect.

3.12. Publicity

This is an even-handed clause which is designed to prevent one party from stealing a publicity march on the other. In practice the two sides would probably agree to a joint release of information.

3.13. Alteration of agreement

This brief clause covers two crucial points. A contract or an amendment to contract can be made orally. It is usual in business transactions to make contracts in writing. The purpose here is to exclude oral variations to the contract. This makes for certainty and ease of administration. Secondly, it is made clear that only properly authorised personnel can enter into contractual commitments. In practice each party should enquire of the other as to which individuals are so authorised. It should be noted, as well, that any amendments, to be effective, are to be signed by both parties. This is distinct from the common arrangement of separate offer and acceptance of contract amendments. The advantage of this approach is that it forces the parties to come to an unequivocal agreement on the content of the admendment before it can be effective and therefore before any work is put in hand.

3.14. Termination

Under contract law, serious breach (which we must assume equates to fundamental breach) entitles the innocent side to summarily terminate the contract. In this clause the defaulting party is given 60 days in which to put the problem right before the contract can be terminated. Interestingly the word 'may' (terminate) is used rather than 'will', giving the buyer the discretion to continue with the contract even if the problem has not been rectified. The possibility of termination, or notification of serious breach, are weighty matters and it should be noted that advice of either must be given in writing.

Depending on the nature of the work under the contract, the subcontractor may ask the main contractor to consider including words placing an obligation on the main contractor to minimise the direct or consequential effect of the breach and to render assistance to the subcontractor, at his cost, as necessary in correcting the problems.

The reference to the right of termination not affecting (referred to as 'without prejudice') other rights is there to convey that, for example, the obligation on the subcontractor to hold the main contractor's information in confidence does not disappear because of the termination. These termination clauses sometimes end by saying'...without prejudice to any other right or remedy'. This has the more general effect of saying that termination does not affect other rights accrued under the contract or under the law at large.

3.15. Notices

Again this emphasises the formal nature of the contract in saying that matters relating to it must be committed to writing. It is also there to assist in contract administration.

3.16. The agreement

a) Most importantly this clause explains that only the contract and its attachments constitute the entire arrangement between the parties. Read in conjunction with clause 3.13 the effect is to say, for example, that a written agreement between the two chief engineers in relation to the work would not be enforceable unless both were nominated as authorised representatives and that their agreement had been incorporated in the contract.

b) This standard statement is included to eliminate any possible doubt of the appropriate jurisdiction in the event of a dispute arising over the contract.

④ Summary

The questions in the mind of the subcontractor in reviewing the offered subcontract are:

  a) Is the type of subcontract offered appropriate to the work?

  b) Is it adequately even-handed?

  c) Are the conditions acceptable?

  d) To what degree of risk am I exposed if I accepted conditions which are not favourable to me?

  e) To what extent will I be successful if I seek to negotiate some of the conditions with the main contractor, and will this in itself cause disaffection in him?

  f) If I choose to comment, to what level of detail should I go?

These questions are largely intertwined and commercial judgement will dictate the answers. In this case the answer to question (*a*) is probably no. The contract is stated to be for the provision of engineering consultancy and yet most of the document alludes to the supply of goods. This clearly is inappropriate, but on the other hand if the clauses cannot in practice apply they cannot cause a problem or expose a risk and so perhaps it is better not to object simply on a matter of principle.

Naturally enough a contract prepared by the buyer will be most favourable to him and typically the subcontractor will accept this principle. Subject to one or two changes only, the offered document is reasonably fair.

The conditions are mostly acceptable although there is considerable risk to the subcontractor if he acknowledges all the main contractor's obligations to satisfy the client and accepts that he could be liable for the consequences of breach of the main contract. After all, the consultancy might in straightforward value terms be worth only a pittance compared to the value of the main contract. However, if the individuals who will provide the consultancy are top class or it is known that the area in which the main contractor needs help is not critical or key to the end-product then perhaps the degree of risk exposure is very small. On balance such clauses are ones to which the subcontractor should seek some revision. The comment regarding prompt payment following rectification of a defective invoice is also worth pursuing. Cash-flow is all-important and it is unacceptable to leave the main contractor with the contractual right to delay payment indefinitely. All points of clarity, such as the question relating to IPR, should be resolved.

The question of success or failure, disaffection or not in seeking to negotiate contract conditions, depends largely on custom and practice, the respective positions and strengths of the two sides and, most importantly, the personal relationship between the two negotiators.

As to the level of detail into which comments on the conditions should go, all the same principles as above apply. For example, the document is less than perfect in so far as it switches between alphabetic and numeric subparagraph nomenclature, but this hardly matters since clarity is not compromised and the document is consistent throughout.

III

● ● ● ● **C** **A** **S** **E** ● **S** **T** **U** **D** **Y** ● ●

Widgets

1. **Scenario**

2. **Contract clauses**

3. **Questions**

4. **Analysis**

① Scenario

1. Spiffo Controls Ltd has accepted a contract from Windup Rotary Ltd to supply 300 mini widgets. The contract was accepted on 1 April with delivery to be complete by 30 October.

2. The contract provides for payment to be made in three stages: 25% in May, 50% in September and 25% in November.

3. The contract contains clauses on delivery, acceptance, risk, property and warranty.

The following sequence of events takes place:

a) Work is put in hand by Spiffo.

b) Windup computer on the blink and first payment to Spiffo is delayed to 1 July.

c) 24 June Spiffo advise 1st batch complete.

d) Windup collect 1st batch, all of which fail acceptance test on 3 July.

e) Windup return 1st batch, 10% of which are damaged in transit.

f) Spiffo deliver 2nd batch on the 28 August. Windup reject the delivery as no prior notice was given.

g) Windup withhold September stage-payment.

h) 28 September Spiffo advise that 90% of the first batch and 100% of the 2nd batch are ready.

i) Windup agree to take delivery of 90% 1st batch and 100% 2nd batch on 30 September although they are not happy about a claim from Spiffo that Windup are responsible for the 10% damaged in transit. Spiffo deliver and the goods pass acceptance test on 18 October.

j) In October Windup make the 2nd stage-payment.

k) 28 October Spiffo advises that 3rd batch is ready. Windup refuses to take delivery until the 10% of 1st batch is delivered. Spiffo advised by Windup that final payment will be withheld.

l) 14 November Windup rejects 5% of the 1st batch under warranty.

② Contract clauses

Delivery

The seller shall deliver the articles to the Buyer's works

at Stockton-over-Puddle. Delivery shall be made in three equal batches according to the following schedule:

Batch 1: Friday 26 June
Batch 2: Friday 28 August
Batch 3: Friday 30 September

Delivery is required in the week leading up to the above-mentioned dates and seller is required to give the Buyer 24 hours notice of consignment. The Buyer shall be under no obligation to accept articles outside of the required delivery or if no notice of consignment is received.

Acceptance

The articles shall be deemed accepted following testing by the Buyer in accordance with the agreed acceptance test plan.

Risk

Risk in the goods shall pass to the Buyer following acceptance.

Property

Property in the goods shall pass to the Buyer on delivery unless the contract provides for interim payments in which event property shall pass to the Buyer during the course of manufacture.

Warranty

The Seller shall warrant the goods against defective materials or workmanship for 30 days. Property and risk in goods rejected under warranty shall revert to the Seller who shall replace the defective goods at his own cost with conforming goods within 15 days.

3 Questions

1. Who is responsible for the 10% first batch damaged in transit bearing in mind that Windup collected and returned the goods?

2. The first payment occurs after Spiffo's first attempt to make the initial delivery. Who owns the first batch before and immediately after the acceptance test is failed?

3. Are Windup right to refuse delivery of the second batch and can Spiffo claim against Windup for the repeat delivery cost?

4. Does Windup have the right to withhold the September stage payment?

5. Should Windup have taken redelivery of 90% first batch?

6. Do Windup have the right to refuse the third batch?

7. Are Spiffo obliged to replace the 5% warranty rejects?

8. By November Windup only have 185 working widgets against their requirement for 300 by the end of October. Spiffo have only 75% of their money, 100 widgets made that they cannot deliver and a warranty claim as well. What should Windup and Spiffo do to resolve their mutual problem?

4 Analysis

1. Spiffo is undoubtedly responsible for the 10% 1st batch damaged in transit after rejection by Windup. The contract clearly says that risk does not pass to the Buyer until acceptance has taken place. Spiffo could argue that Windup were at fault in collecting the goods in the first place and, as such, how could Spiffo know that the fault causing rejection was not caused by or during transportation? They could also argue that Windup should have advised Spiffo of the acceptance-test failure and required Spiffo to collect the rejected goods. In the former case the contract says that Spiffo are responsible for delivery, in the latter the contract is silent on who transports rejected goods. Neither of these facts would affect the outcome, Spiffo are bound to repair or replace the damaged goods. It is to be hoped that, in signing a contract that includes the responsibility for delivery and in which risk passes to the Buyer at a late stage, Spiffo carry the necessary insurance cover against loss or damage in transit. A point to be noted is that it is prudent to check that the insurance does cover rejected goods as well as conforming items.

2. On the question of ownership of the 1st batch before and immediately after the first delivery attempt, a deficiency in the wording of the delivery clause is discovered. Property, i.e. ownership, passes to the buyer on delivery but the delivery clause only refers to the timing on physical delivery and the obligation on the seller to effect transportation. Thus it is not clear whether delivery, for the purpose of identifying the passage of property, is physical arrival of the goods or their acceptance. However, the point becomes academic as property passes to the buyer during manufacture in view of the provision in the contract for interim payments. The fact that the initial payment was delayed is not relevant as the contract *provides* for interim payments, i.e. it does not say that interim payments have actually to be made for property to pass at the early stage.

3. The contract gives Windup the right to refuse delivery if no notice is given. Spiffo therefore have no claim in respect of repeat delivery costs. In practice Spiffo would be wise to try and find out more about the refusal to take delivery. On the face of it Windup needed delivery at specific points in time,

presumably to meet some production programme. If the requirement for prior notice was included for no more than the convenience of Windup then perhaps there is something more
behind the refusal than meets the eye. At this stage in the
sequence of events Spiffo would be right to be suspicious and
should try to find out more.

4. Without seeing the payments clause in detail it is not
possible to say whether Windup had the right to withhold the
September payment. Interim payments tend to be discretionary
and so Windup may have had the right. Windup would argue
that by the beginning of September Spiffo were seriously in
default of contract. The 1st batch all failed acceptance test and
no notice was given for delivery of the 2nd batch and thus
payment should be suspended. Windup would argue that
Windup's collection and subsequent return of the 1st batch may
have contributed to the problem, that they had not said that the
rejected articles would not be replaced and that the 2nd batch
had been refused on a technicality. In any event interim
payments were a fundamental feature on the contract and should
not have been withheld.

5. There was nothing to stop Windup from taking redelivery of 90% of the 1st batch. They would have been entitled
not to take delivery as the contract clearly said delivery in
batches. This partial delivery could not be used as evidence by
Spiffo that Windup were acknowledging reponsibility for the 10%
damaged in transit.

6. Windup should not really have refused the third batch.
Some sympathy lies with Spiffo. They have had payment delayed
or withheld three times and yet they have successfully delivered 90% of batch 1, 100% of batch 2 and were ready to deliver
100% of batch 3. From their point of view Windup has been a
fickle customer. Payments have been delayed or withheld,
deliveries have been refused, articles that should have been delivered have been collected and all these have been virtually
random events. On top of this the early manufacturing problems
– if that was the reason causing acceptance failure of the 1st
batch – have gone away and Spiffo must be desperate to finish
the order, take the money and run.

7. The problem of warranty is not clear as the warranty

clause is not precise as to when the warranty period commences. Windup have advised the warranty claim within thirty days of the goods passing acceptance test but beyond thirty days following physical delivery at their works. Windup would argue that, as the contract specifically calls for an acceptance test, then clearly before that test has been conducted it is impossible to say whether the articles are good or bad and thus warranty must commence following completion of the acceptance test. Spiffo might say that as they have no knowledge or control over when Windup would conduct the test it is only reasonable for the warranty to commence on physical delivery, otherwise Spiffo could carry an indeterminate liability. Windup could respond by saying that the warranty period in itself is irrelevant as clearly the goods are not of merchantable quality and/or reasonably fit for their purpose. Only a court could decide that question and neither party would want to go that far at this stage. In negotiating the contract Spiffo should have tried to close off that risk by agreeing the specific warranty on the condition that it is the only warranty, i.e. those implied by the Sale of Goods Act are excluded.

8. Although the greater sympathy might be with Spiffo, the upper hand is held by Windup. Spiffo must start with the assumption that Windup are not capriciously or maliciously being difficult. On that basis, Spiffo should establish what is the maximum that Windup would expect from them such that, if Spiffo yielded all those things, Windup would have no excuse for not restoring payments and taking deliveries. The maximum that Windup could want is:

a) An agreement from Spiffo to replace the 10% 1st batch damaged in transit by a specific date at no cost to Windup.

b) An acceptance of the warranty claim, giving Spiffo a commitment to replace the defective goods quickly, if not within the 15 days, at no cost to Windup.

c) Delivery of the 3rd batch.

d) Agreement that warranty commences at successful completion of the acceptance test, rather than on physical delivery.

e) Compensation from Spiffo for delays caused by non-delivery, late delivery and rejected goods.

Assuming that the 3rd batch will not fail the acceptance test, the 10% damaged units and 5% warranty claim actually means that Spiffo only have to replace 15 units out of 300. This may be an acceptable penalty in order to get the final 25% of their money. Provided Windup agree not to delay final payment, Spiffo should make an offer to accept liability to replace the 15 units such that delivery of the 3rd batch can go ahead. If Windup insist on a retention until at least the 10% of the first batch have been delivered and accepted, Spiffo should try to negotiate a much lesser sum than the 25% owed to them. Spiffo may be able to accept that warranty starts from completion of acceptance testing provided Windup agree to do the tests within a specified maximum period following delivery.

It is in Spiffo's best interests to put this type of solution – phrased to say that this is in full and final settlement of their obligations – forward to Windup before they think of, or make a claim for compensation. Such a claim, were it to be sizeable, could lead ultimately to the unthinkable – a law-suit.

The major lessons to be learned are that the contract clauses must be precise and deal with all possible eventualities. It is important that seemingly minor obligations under the contract – the requirement to give notice of delivery – are not ignored in practice. Additionally the supplier must keep in mind the consequences (cash-flow, reputation, level of stock, insurance, legal eventualities) of not performing the contract exactly as required.

II

Explanation of terms

Acceptance
Point at which a buyer has acknowledged and accepted that delivered goods conform with the contract and thus the point at which the buyer loses his right to reject the goods.

Acts of God
Literally, acts of god – storm, flood, etc., which may be included in the contract under a *force majeure* or excusable-delays clause.

Added value
The enhancement of the value of materials through the application of manufacturing processes.

Amendment
An agreed variation to a contract that in effect creates a new contract and which therefore, to be binding, must satisfy all the criteria of a contract.

Application date
Date on which a patent application is filed at the patent office.

Articles of association
The document designed to regulate the internal affairs of a registered company.

BEAMA
British Electrotechnical and Allied Manufacturer's Association.

Breach
Failure to comply with a term of a contract.

Budgetary estimates
See Estimate.

Cash-flow
Analysis of the rate of receipt of monies with the rate of disbursement of monies.

CBI
Confederation of British Industry.

Civil law
That part of the law comprised of public law and private law but excluding criminal law.

Commercial awareness
The ability to take the broad view across all of the objectives and operations of the business and to understand and implement the vital elements.

Common law
Law developed since antiquity through judicial precedent.

Company secretary
The chief administration officer of a registered company appointed by the board of directors.

Condition
A fundamental term of a contract, breach of which entitles the injured party only to sue for damages.

Confidentiality agreement
An agreement between two or more parties, each promising to the other(s) that proprietary information exchanged between the parties for a specified purpose will be treated by all parties as confidential and afforded appropriate safeguards against unauthorised disclosure.

Consideration
The legal word for the money (or other thing of value) paid in a contract.

Contingency
A provision made by the supplier within the price for a possible event for which it is impossible to estimate accurately.

Contract
A legally binding exchange of promises.

Copyright
The concept of the concrete expression of an idea being protected against reproduction.

Corporation
A single individual or group of individuals carrying on a business

but possessing legal entity separate from that of the individual(s).

Cost
The break-even point between a financial loss and surplus (profit).

Cost plus
Short-form name for the type of contract in which the seller is reimbursed his actual incurred costs plus a fixed or percentage fee.

Counter-offer
An alternative proposal made in response to an offer of contract.

Criminal law
A division of public law covering forms of conduct for which the state prescribes punishment.

Current assets
The sum of raw materials, work in progress, finished goods, cash in hand/bank, debts.

Damages
Compensation for injury, loss or damage.

Default
Failure to discharge an obligation under the contract.

Deliveries
Value rate at which a contract is performed.

Direct costs
Costs directly and solely attributable to individual contracts, products or projects.

Directors
Persons chosen by company shareholders to conduct and manage the company's affairs.

Dividends
Earned profits divided between the shareholders.

Estimate
Indication of cost or price: generally accurate to ±15% only.

Estimating allowance
A provision made by the supplier within the price for a known event for which it is impossible to estimate accurately.

Equity
Another name for share capital.

Exclusive licence

A form of licence permitting only the licensor and licensee to use and exploit the subject's intellectual property.

Excusable delays

Another expression for *force majeure*.

Express warranty

A specific clause in a contract under which the seller promises for a specific time to remedy defects discovered after acceptance of goods by the buyer.

Filing

Lodging a patent application at the patent office.

Firm price

Price not variable for any reason other than a change to the work content or duration.

Fixed assets

Expenditure on land, buildings, plant and machinery.

Fixed price

Price, the final value of which is set by reference to some variable factor.

Force majeure

A provision within a contract that excuses the seller for late delivery if the cause was beyond his control.

Goodwill

The realisable value flowing from a company's good reputation.

Gross order book

Value of all orders held.

HMG

Her Majesty's Government.

Implied warranty

An obligation placed upon the seller by a court or under the Sale of Goods Act to remedy defects discovered after acceptance by the buyer even though no such obligation was specifically written into the contract.

Incentive contract

Contract in which the price is determined by reference to degree of success in meeting or bettering specified targets of cost and/ or time and/or performance.

Indirect costs

Costs not attributable to individual products or projects.

Industrial property rights
Synonymous with Intellectual Property Rights.

Instruction to proceed
A form of offer of contract used in urgent circumstances which if accepted creates a contract.

Intellectual property
The ideas, inventions, designs, know-how and reputation unique and belonging to a particular company.

Intellectual property rights
The rights arising from contract, law or statute to protect and exploit intellectual property.

Intent to proceed
An indication only of intention to place an contract: not capable to acceptance and having no contractual significance.

Investment
Earned profit retained for purposes other than direct reinvestment in the business.

Joint stock company
An organisation consisting of persons who contribute money to a common stock which is employed in some trade or business, and who share the profit or loss arising.

Know-how
Particular knowledge and skills peculiar to an individual or company.

Licence
The grant of one party to another of the right to use or exploit intellectual property in return for payment. The grant of the right does not transfer ownership of the intellectual property.

Licence fee
A payment made by a licensee to a licensor not based directly on the value of the licence of the licensee.

Licensee
The party to whom a licence is granted.

Licensor
The party granting a licence.

Limited company
A company in which the liability of shareholders is limited to the amount unpaid on their shares in the event of company wind-up.

Liquidated damages

An arrangement in a contract whereby the seller agrees to pay the buyer specific sums of money by way of compensation for, but only in the event of, late delivery.

Loan capital

Capital borrowed from a bank or other lender.

Memorandum of association

The document designed to regulate the external affairs of a registered company.

Negligence

Breach of the duty of care, a duty which everybody has under common law.

Net order book

Value of gross order book less value of deliveries made.

Non-transferable licence

A licence which the licensee cannot pass on to a third party.

Output

See Deliveries

Overheads

See Indirect costs

Ownership

Having legal title to goods.

Partnership

Relationship between two or more persons carrying on a business with a view to profit.

Patent

Statutory protection given in respect of a technical concept which grants monopoly rights to the inventor in return for disclosure of the idea.

Patent search

A search to determine the novelty or otherwise of the subject concept.

Possession

The actual physical holding of goods.

Price

The sum of money which the customer pays.

Private law

Legal relationship between individuals or groups of individuals.

Private limited company (Ltd)
Similar to a public limited company but having no right to offer shares to the general public.

Profit
The difference between the cost of a job and the price which the customer will pay.

Profit and loss account
An annual picture of the company's financial performance.

Property
In simple terms this has the same meaning as Title, q.v.

Proprietary
Belonging exclusively and uniquely to a particular company.

Public law
Legal relationship between individuals or groups of individuals.

Public limited company (Plc)
A limited company having a minimum £50,000 share capital that is permitted to offer shares to the general public.

Quotations
Definite price offer upon which the customer can rely. Capable of formal acceptance to create a contract provided acceptance is given within the period of validity.

Registered company
A company registered under and in compliance with the United Kingdom Companies Act.

Rejection
Buyer's right not to accept goods if within a specified or reasonable time he has advised the seller that the goods do not conform to the requirements of the contract.

Reserves
Earned profit reinvested in the business to generate growth and development.

Risk (1)
Liability for loss of damage to goods.

Risk (2)
The exposure to possibilities of commercial, financial and legal damage arising from carrying on a business operation.

Royalty
A payment made by a licensee to a licensor based on a proportion of the value of sales made by the licensee.

Share

The proportion of the stock to which each member (share-holder) is entitled by virtue of his investment.

Share capital

Capital belonging to the person(s) setting up a business.

Sole licence

A form of licence permitting only the licensee to use and exploit the subject intellectual property, i.e. to the exclusion of all others including the licensor.

Sole trader

A single individual taking all the risks and benefits of carrying on business.

Standard conditions

Contract conditions established by individual companies, sectors of industry or between regular contracting parties that serve to reduce the time and cost of preparing and negotiating contracts.

Statute

An Act of Parliament imposing requirements or obligations.

Stock

The capital of a company.

Terms

All of the requirements and obligations upon parties to a contract which, together and with the benefits that each enjoys under the contract, constitutes the entire contract.

Time is of the essence

A contractual phrase meaning that making delivery at specific times is of fundamental importance to the buyer.

Title

Legal expression meaning to enjoy ownership of goods and the right to utilise, modify, sell or otherwise dispose of without encumberance.

Trade marks

An identifiable symbol so particularly associated with a specific company that others can be prevented from using it.

Trade names

Similar to trade marks but relating to actual names.

Trade secrets

A generic expression encompassing know-how, inventions and commercially sensitive information.

Trading identities

Synonymous with trade names.

Unlimited company

A company in which the liability of share holders is unlimited in the event of company wind-up.

Validity period

Period of time within which a quotation must be accepted if acceptance is to create a contract.

VOP

Variation of Price: an arrangement whereby the contract price is varied according to the movement of some variable parameter such as inflation.

Warranty

A non-fundamental term of a contract, breach of which entitles the injured party only to sue for damages.